THE
GREAT
PIVOT

Creating Meaningful Work
to Build a Sustainable Future

JUSTINE BURT

Interior design by Tabitha Lahr

Published 2019 by MP Publishing
Printed in the United States of America
ISBN: 978-1-935994-34-3
E-ISBN: 978-1-935994-33-6
Library of Congress Control Number: 2019902752

For information, email: thegreatpivot@gmail.com

CONTENTS

CONTENTS

Dedicated to Chris and Matthew

"*I want to be with people who submerge*
in the task, who go into the fields to harvest
and work in a row and pass the bags along,
who are not parlor generals and field deserters
but move in a common rhythm
when the food must come in or the fire be put out."

—Marge Piercy
"To be of use"

LENGTHENING OUR TIME HORIZONS:

INTRODUCTION TO THE GREAT PIVOT

n 2017, an epiphany hit me while touring a gothic cathedral on vacation in Northern Spain. As my family and I learned the cathedral's history, I realized part of the reason we're not building a sustainable future fast enough in the United States, and it has to do with our short-term time horizon, which lies in marked contrast to the much-longer time horizon of the cathedral builders.

The beautiful Santa Maria Cathedral is perched atop a hill overlooking the town square and has been the pride of Vitoria-Gasteiz since construction was completed in 1437. Think about how the original architects, patrons, and workers of the cathedral saw their role in history: they knew they would not be around to see its completion, yet they worked together to create something awe-inspiring and majestic for future generations. It would be their legacy.

Six centuries later, the cathedral started showing its age. In the 1960s, the water that had seeped into the foundation for centuries had begun to compromise the building's structural integrity. Interior gothic arches started to lean visibly, and a large stone block fell out of the ceiling onto the altar, smashing it. Concerned about the long-term viability of their architectural treasure, the community rallied to secure funding for renovations and began this careful work. Teams reinforced the foundation, braced the arches, and rebuilt the altar. The community wanted Santa Maria Cathedral to last for several more centuries.

Walking through this masterpiece, with its detailed stonework and stained-glass windows, I mused over how people in the 14th century had considered deeply their impact on the future, and how the town over the last 50 years had been working diligently to reinforce the cathedral as their legacy and gift to distant generations. But what are we in the United States doing to benefit people centuries from now? Will the inheritance of future generations be climate change, mass extinction, and a degraded natural world?

We have to do more to forge a different path, I thought. If we are to lengthen our time horizon and claim our place in history, we must devote more resources to steering society in a more sustainable direction.

We are building renewable energy systems, retrofitting buildings for energy efficiency, creating a circular economy, and restoring natural systems—it's just not happening fast enough to get ahead of the destruction. We need to step up our efforts. We must invest in more resilient systems that can bounce back after a natural disaster or other shock. That should be our legacy and our gift to future generations. And it all starts by pulling back our perspective and taking a longer view.

There are two types of people who can make this happen: those who want to do meaningful work and policymakers in a position to facilitate meaningful job creation. This book is written for both of them. Together, these two groups can create stable, middle-class jobs (that cannot be outsourced or automated) in the private, non-profit, and government sectors, which in turn will give millions of lives a sense of purpose and will stabilize our relationship with the planet that we rely on for our survival.

ORIGIN OF THE GREAT PIVOT

Those moments wondering how future generations will look back on our time are part of the origin of the title *The Great Pivot*. Environmental activist Joanna Macy coined the term "the Great Turning" to express the idea that we need to turn away from the linear, non-renewable, extractive, polluting trajectory we are currently on. We must ask ourselves, do we want future generations to look back at our time as the Great Unraveling, or should we instead choose a more sustainable path so that our grandchildren see this era as the Great Turning? We are at this crossroads, and we have the power to determine our destiny.

The other element of the title borrows the term "pivot" from the startup world. When startups launch, they work from a draft of their business model. As the team builds the company and attempts to validate their business model, they may at some point realize that if they do things a little differently, they might have a financially sustainable business on their hands—and this is when they pivot. By making some changes in their offering or how they run the business, the pivot will allow them to realize a replicable and scalable business model. Thus, combining the terms "great" from the Great Turning with "pivot" from the world of startups, we arrive at "the Great Pivot."

BASIC ELEMENTS OF THE GREAT PIVOT

Our society's current business model is not sustainable. The conventional ways we generate energy, transport people and goods, extract materials and manufacture products, grow and distribute food, and treat natural systems cannot be maintained. We need to pivot. Doing so will:

1. Address the crisis in the world of work
2. Scale up the building of a sustainable future
3. Create meaningful jobs that people so desire
4. Generate more sustainability projects in which the financial sector can invest

By making the Great Pivot, we will establish the legacy of a sustainable society for future generations and, within our lifetimes, create meaningful work for millions of people.

The term "meaningful work" refers to jobs that bestow a sense of purpose on the people doing them. Some may call the jobs detailed in this book "green jobs," but *The Great Pivot* seeks to put a finer point on the term by calling them "meaningful jobs." Whether we call them meaningful or green, we have much work ahead of us to build a sustainable future, one that will allow us to live in balance with our environment.

There's a story about three people breaking rocks by the roadside. When a passerby asks what the first person is doing, he responds, "I'm breaking rocks."

When asked, the second one says, "I'm feeding my family."

The third one says, "I'm building a cathedral."

A sustainable future needs us to make it happen: generations not yet born are depending on us. By embracing and implementing the vision and blueprint of the Great Pivot, we will serve as the cathedral builders that future generations need us to be.

PART I:
THE PROJECTS

1

BUILDING A WORLD OF

MEANINGFUL WORK

"Bring all of you to work."
—Andre Delbecq, former Dean of
Business Administration, Santa Clara University

Caterpillar engineers in Decatur, Illinois, are developing mammoth, self-driving, coal-hauling trucks for the mining industry. The 850,000-pound vehicles will not have drivers or even remote operators. By eliminating drivers, the coal industry will be able to cut costs and continue to compete with other forms of energy for a little while longer. Despite President Trump's promises to dismantle the Obama administration's climate change efforts and bring back coal jobs, the coal industry is eliminating jobs partly because of automation.[1]

UPS truck driver Paul doesn't worry about losing his job to automation, but other people at his company do. "My job's secure, but the people in the warehouse moving packages around, their jobs are going away. We're already seeing it. Robots can't do my job delivering packages, though, at least not yet. It won't be [possible] until well after I'm gone."

People like Paul sense a big shift underway. Besides automation, other elements such as outsourcing, the gig economy, and low levels of employee engagement are all contributing to a crisis in the world of work. This sense of unease creates cognitive dissonance when the news reports rosy economic indicators. In December 2018, the official unemployment statistic informed us that only 3.9% of Americans of working age were unemployed.[2] The reality, however, is much starker: there are actually 37 million people of prime working age, between the ages of 25 and 64, who are Not in the Labor Force (NILF). Officially, our economy is at full employment, with millions of jobs begging for workers; less well known is that tens of millions of people are not working and that the majority of those who are working are not engaged at work.

There is a yawning gap between the employment we have and that which we deserve: stable, engaging work that fills our lives with purpose. To start to close this gap, we need to better understand the nature of the crisis.

The U.S. Department of Labor's Bureau of Labor Statistics (BLS) generates unemployment numbers each month. The most widely touted labor statistic is the U-3 unemployment rate. Reputable news outlets in October 2018 celebrated a historically low 3.7% U-3 unemployment rate in the previous month. Articles reported that, officially, there were more job openings than there were people looking for work. The Labor Department said there were 7.14 million job openings, the highest on record since December 2000, and 6.2 million unemployed people.[3]

Looking at a 3.7% unemployment number, one would think the economy has lifted all boats and everyone who wants a job has one. Why, then, is there such a discrepancy between the official numbers and the lived reality of millions of Americans?

THE MISLEADING U-3 UNEMPLOYMENT STATISTIC

When the news reports the U-3 unemployment rate, they are referring to a very specific subsection of people. To count as unemployed, one needs to have actively looked for work in the previous four weeks. Actively looking for work consists of any of the following activities:

- Contacting an employer directly or having a job interview

- Contacting a public or private employment agency
- Contacting friends or relatives
- Contacting a school or university employment center
- Submitting resumes or filling out applications
- Placing or answering job advertisements
- Checking union or professional registers
- Some other means of active job search

Every month, the BLS surveys 60,000 households to determine the residents' work statuses and then place approximately 110,000 people into one of three main categories: employed, unemployed, or NILF. The monthly survey interviews NILF persons about their desire for work, the reasons why they have not looked for work in the last four weeks, their prior job searches, and their availability. These questions include the following:

- Do you currently want a job, either full or part time?
- What is the main reason you were not looking for work during the last 4 weeks?
- Did you look for work at any time during the last 12 months?
- Last week, could you have started a job if one had been offered?

According to the BLS, in the second quarter of 2018, among the civilian non-institutionalized population between the ages of 25 and 64, 126.8 million people were employed and 3.9 million people were unemployed and actively looking for work, while 37.5 million people were NILF. Keep in mind that the three groupings of employed, unemployed, and NILF do not include those in the military, prison, long-term nursing care, or mental health facilities. The 1.29 million active duty military personnel[4] are not counted among those employed, and the 2.3 million people in prison do not count as unemployed or NILF.

So who does count as NILF? For those of prime working age, NILF includes people who:

- Are ill or disabled
- Could not find work
- Are taking care of home/family

- Are going to school
- Are retired
- Other

The BLS has a NILF category for discouraged workers, meaning people who are not currently looking for work for one of the following reasons:

- They believe no job is available to them in their line of work or area
- They had previously been unable to find work
- They lack the necessary schooling, training, skills, or experience
- Employers think they are too young or too old
- They face some other type of discrimination

One particular labor statistic that may cause concern has to do with males in the U.S. In 2018, among men aged 25–64, 67.8 million were employed, 2.1 million were unemployed, and 12.5 million were NILF. The fact that these 12.5 million men—9.0 million of whom are white— hear that the 3.7% unemployment rate is the lowest in a generation and that everyone who wants a job has one is disheartening. Because our society expects men of prime working age to be contributing members of society, such a large number of people without paid employment may be causing instability in our society, which manifests itself in myriad ways. To fix this, we must do more to create meaningful jobs for people not in the labor force, not only so they can support themselves but also to bolster their sense of fulfillment and purpose.

Another large chunk of people who could contribute more to the economy are those who work part-time but who would like to work full-time. BLS surveys in the second quarter of 2018 identified about one million people between the ages of 25 and 64 who fit this description. These folks are included in the U-6 unemployment rate, a larger category that includes the U-3 unemployment rate plus:

- Those neither working nor looking for work but who want

and are available for a job and have been looking for work
sometime in the past 12 months
- Discouraged workers who have given a job-market-related
  reason for not currently looking for work
- Persons employed part-time who want and are available for
  full-time work but have had to settle for a part-time schedule[6]

Of the 37.5 million people NILF, some feel discouraged after numerous unsuccessful job application submittals, while others would like to
re-enter the work world after taking time off to care for a family member.
All of this is to say that millions of people looking for more work may
be available.

OUTSOURCING TRENDS

One factor that has raised anxiety about the work world over the last few
decades has been the trend of outsourcing work to other countries. Since
2000, the U.S. has lost five million manufacturing jobs to outsourcing.[7]
As U.S. manufacturing companies looked for ways to reduce labor costs,
they moved operations to Mexico in the 1990s and later to Asia. Clearly,
labor is less expensive in countries where employers are not required
to pay a living wage, vacation pay, healthcare, childcare, insurance, or
retirement benefits.

The cost differential for labor in different countries is striking. A
friend who works at a U.S.-based medical device manufacturing company shared that the hourly rate of labor for subcontractors in China as
compared to the U.S. is not even close. "In the U.S.," he said, "we paid
$30/hour, and in China we pay $3/hour for the same work, and that
price includes delivery."

Adding outsourcing insult to outsourcing injury for U.S. workers
is what often happens right before a U.S. worker is laid off: at times, a
condition of receiving a severance package is that the person being laid
off must train their replacement who resides in another country.

Around 2009, though, some of those outsourced manufacturing
jobs started coming back. Figure 1 shows the trend line for manufacturing in the U.S. since 1980.

FIGURE 1: NUMBER OF MANUFACTURING EMPLOYEES IN THE U.S.

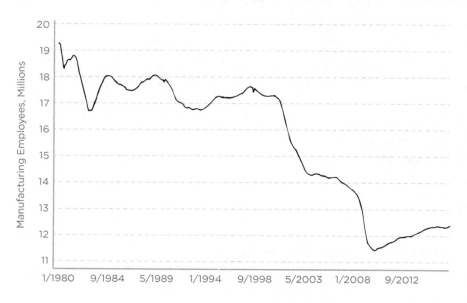

Source: Bureau of Labor Statistics. seasonallv adiusted

As some of those manufacturing jobs were "reshored," the jobs that returned were not the same ones that had left. With advancements in manufacturing automation, the 1,000 manufacturing jobs that had been outsourced to China might have come back as ten jobs that now require workers to have knowledge of robotics, electronics, and software programming. For good reason, the growth of automation in manufacturing over the past decade has given rise to a new level of anxiety about the work world.

THE UPSIDE OF INCREASING AUTOMATION

Articles about robots taking our jobs have proliferated the last several years, stoking fears. Not knowing the degree to which automation will replace humans is a scary prospect, but we should actually see this trend as an opportunity. Robots are better positioned to do the types of tedious or dangerous work that cause humans repetitive stress and other injuries. Looking at the bright side, automation offers many benefits, foremost among them that it frees up the U.S. labor force to undertake more compelling, fulfilling types of work.

To make up your own mind about whether the trend toward automation poses a blessing or a curse, take a look at Figure 2 from McKinsey & Company. This graphic explains which types of work are most vulnerable.

FIGURE 2: ACTIVITIES MOST SUSCEPTIBLE TO AUTOMATION

Technical Feasibility, % of Time Spent on Activities That Can Be Automated by Adapting Currently Demonstrated Technology

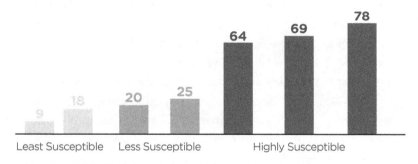

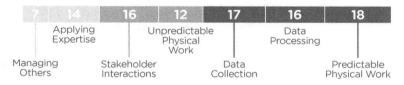

Source: *McKinsey Quarterly*

On the right side of Figure 2, we see that predictable physical work, data processing, and data collection are most susceptible to automation. Less-predictable work where people manage others or apply expertise are less susceptible.

A study about automation McKinsey & Company published in 2017 explained that "very few occupations, less than 5%, consist of activities that can be fully automated." However, it continues, "in about 60%

of occupations, at least one-third of the constituent activities could be automated, implying substantial workplace transformations and changes for all workers."[8]

What if this situation were not a tragedy but rather something to be embraced?

Dennis Mortensen, CEO of New York City-based artificial intelligence (AI) startup x.ai, outlines the upside of automation. The company provides AI personal assistants, and Mortensen believes AI can free us from tedious tasks to do more creative work that we actually enjoy. Consider how Mortensen believes AI could change a salesperson's job.

> "Right now, a sales assistant likely spends a lot of time doing things that could be automated: prospecting for and qualifying leads, sending follow-up emails, updating Salesforce, building reports, etc. Once all that's taken over by intelligent machine agents, what's left for you as a salesperson? It's the emotional and creative stuff. You'll spend your day building relationships and serving your clients with creative solutions to their problems. By freeing you from the mundane tasks you used to have to do, often grudgingly, AI will let you focus on things that form the core of your job: the stuff that only you, a human, can do."[9]

Automation has freed humans from drudgery many times in the past. Each time the replacement happened, people grew nervous, to put it lightly, that machines would make humans obsolete, but people then moved on to different kinds of work.

In 1698, Englishman Thomas Savery built a steam pump that he called "the Miner's Friend," as it was used to pump water out of coal mines. Savery was not the first to build a steam pump—people had been building and improving upon them for centuries—but his design released people from having to haul water out of the mines by hand.[10]

In 1793, Eli Whitney invented the cotton gin, prior to which people separated cotton fibers from their seeds by hand. The cotton gin allowed a jump in human productivity when it replaced people in this tedious and painful task. The cotton gin is widely acknowledged to be one of the key technologies that drove the Industrial Revolution.[11]

For decades, people had been using wheelbarrows and rakes to spread asphalt for roads. Then, in the 1930s, Harry Barber created the mechanical asphalt paver, an advancement that allowed road builders to distribute and spread asphalt faster and with greater uniformity.[12]

Looking back, do we wish people were still hauling water out of coal mines with buckets, manually separating cotton fibers from the seeds, or spreading asphalt with wheelbarrows and rakes? Of course not, and we could approach the coming automation revolution in the same spirit. What we must determine is, what do we want to be doing in the future, and what should the future world of work look like?

The transition to this different, future work world may be painful. The prospect of having to learn new skills can be daunting, but if we are open to being retrained, we may end up doing work that we enjoy more than our current jobs.

THE INSTABILITY OF THE GROWING GIG ECONOMY

Compounding human anxiety about outsourcing and automation is the growing gig economy. People who piece together their livelihood with a few different part-time jobs are part of the "gig economy." Gigs may involve working eight hours per week for three months, a one-week stint, or something as short as a half-day. This type of temporary work that offers no benefits may also be referred to as independent contracting, temping, freelancing, consulting, contingent work, and alternative work. Whatever we call it, the ranks of those working without stable employment or benefits is large and growing.

The Bureau of Labor Statistics reported that in May 2017, 3.8% of workers (5.9 million people) held contingent jobs.[13] These contingent workers do not expect their jobs to last or report that their jobs are temporary. Another 6.9% (10.6 million) were working as independent contractors, on-call workers, temporary help agency workers, or for contract firms. These categories describe workers at firms such as Uber, Lyft, Upwork, and Task Rabbit but also businesses in many other corners of the work world.

Some estimates count gig workers as only 0.1% of the economy, while other estimates place the number much higher. The Freelancers Union calculates that 36% (57.3 million) are contract workers.[14] Survey research conducted by economists Lawrence Katz of Harvard University and Alan Krueger at Princeton University finds that "94% of net job growth between 2005 and 2015 was in the alternative work category" and "over 60% was due to the rise of independent contractors, freelancers, and contract company workers."[15]

This growing trend of gig work is only expected to continue. The title of an October 2017 *Forbes* article asks, "Are We Ready for a Workforce That Is 50% Freelance?"

There is a big upside to the gig economy: many gig workers appreciate having a flexible schedule, as they can work when they want. Flexible work arrangements benefit those caring for children or aging parents or those who simply want the freedom to set their own schedules. The downsides of gig work may outweigh the benefits, though: gig work generally does not offer benefits, job security, or protections, and gig workers are often anxious about whether they will be able to make ends meet. Compared to people with full-time jobs, 63% of gig workers dip into their personal savings at least once a month to smooth over gaps, versus 20% of non-freelancers.[16]

LOW LEVELS OF EMPLOYEE ENGAGEMENT

The last of the four factors contributing to a crisis in the world of work is the low level of employee engagement in the United States. Gallup polls 1,500 workers on a regular basis (up until July 2017, surveys took place daily) and finds that only about one-third of U.S. workers are engaged at work.[17] Figure 3 illustrates the levels of employee engagement between 2014 and 2017.

FIGURE 3: U.S. EMPLOYEE ENGAGEMENT

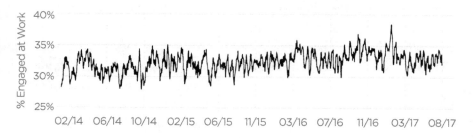

Source: Gallup

Gallup defines engaged employees as those who are involved in, enthusiastic about, and committed to their work and workplace. The twelve "Do you feel that"-type questions Gallup asks in their surveys include:

1. I know what is expected of me at work.
2. In the last seven days, I have received recognition or praise for doing good work.
3. At work, my opinion seems to count.
4. I have a best friend at work.
5. I have the materials and equipment I need to do my work right.
6. My supervisor, or someone at work, seems to care about me as a person.
7. The mission or purpose of my company makes me feel my job is important.
8. In the last six months, someone at work has talked to me about my progress.
9. At work, I have the opportunity to do what I do best every day.
10. There is someone at work who encourages my development.
11. My associates or fellow employees are committed to doing quality work.
12. This last year, I have had opportunities at work to learn and grow.[18]

While many of these employee engagement questions revolve around how supported workers feel at work, the seventh question about the mission or purpose of the company cuts to the heart of meaningful work. If employees support and are excited about the mission of their company, their work will have more meaning for them. Thus, businesses whose work focuses on sustainability will inherently have a strong mission, thereby evoking higher levels of employee engagement.

What is the effect on businesses with higher rates of employee engagement? Gallup compares business units in the top quartile of employee engagement and business units in the bottom quartile and finds those in the top quartile enjoy:

- 41% lower absenteeism
- 24% lower turnover in high-turnover industries
- 59% lower turnover in low-turnover industries
- 70% fewer safety incidents
- 40% fewer quality defects
- 10% higher customer metrics
- 17% higher productivity
- 20% higher sales
- 21% higher profitability[19]

For employers that want to attract the next generation of workers, investments in worker engagement and sustainability may be worth the time and expense. Younger generations of workers demand more from their employers, and because Millennials are the largest generation in the workforce (by 2020 they will make up 50% of employees in the U.S.), they will continue to effect change. They not only seek greater purpose and to work in organizations with strong social and environmental commitments but also want to be involved in developing those commitments.

The 2016 Cone Communications Millennial Employee Engagement Study found that "three-quarters (76%) of Millennials consider a company's social and environmental commitments when deciding where to work, and 88% say their job is more fulfilling when they have opportunities to make a positive impact on social and environmental issues."[20]

Shared mission is a key element that inspires potential hires and can unify an existing work team. If people feel they are a part of a team doing something that they are proud of—something they could not accomplish separately—this sense of collective purpose will make them happier employees. When people have an opportunity to work in green jobs with a team building advanced energy communities, low-carbon mobility systems, and a circular economy, or reducing food waste and restoring nature, this work gives them a feeling of meaning and purpose.

In the private sector, companies with compelling core values, such as business being a driver for social change and environmental improvement, ignite passion and motivation in their employees. Take a look at the mission statements from four sustainability-minded companies with strong employee loyalty in the food, beverage, and consumer product industries.

- **Clif Bar**—"We believe in creating a healthier, more just and sustainable food system."
- **New Belgium Brewing**—"To manifest our love and talent by crafting our customers' favorite brands and proving business can be a force for good."
- **Seventh Generation**—"To inspire a consumer revolution that nurtures the health of the next seven generations."
- **Patagonia**—"Build the best product, cause no unnecessary harm, use business to inspire and implement solutions to the environmental crisis."

These inspirational mission statements help drive each company's culture and engage their workforce. This is evident when I see managers from these companies present at sustainable business conferences. Enthusiasm for their work and their company shines through.

The empowering effect of a strong sense of mission applies to service companies as well as product companies. Any company creating green jobs where employees work to decarbonize our energy system, transform our mobility system, create a circular economy, reduce food waste, or restore nature will be imbued with a feeling of purpose.

JOBS THAT ARE NOT MEANINGFUL

At the opposite end of the spectrum sits a different type of job, for which Anthropology professor David Graeber has a blunt term: "bullshit job." In 2013, Graeber wrote an article in *Strike! Magazine* called "On the Phenomenon of Bullshit Jobs: A Work Rant."[21] Graeber struck a nerve, as the article was shared over one million times. Afterward, hundreds of people sent him tweets describing their misery at work. These tweets formed the basis for his 2018 book *Bullshit Jobs: A Theory*.

In the book, Graeber defines a bullshit job as:

"A form of paid employment that is so completely pointless, unnecessary, or pernicious that even the employee cannot justify its existence even though, as part of the conditions of employment, the employee feels obliged to pretend this is not the case."[22]

It is critical to note that the only person who can make a value judgment about whether a job is pointless, unnecessary, or pernicious is the employee who holds that position. In an interview in *The Economist*, Graeber explained this point, saying:

> "Something like 37–40% of workers, according to surveys, say their jobs make no difference. Insofar as there's anything really radical about the book, it's not to observe that many people feel that way but simply to say we should proceed on the assumption that for the most part, people's self-assessments are largely correct."[23]

An objective litmus test of a bullshit job is whether the disappearance of the job would have a notable effect. Entire classes of jobs, if they were to disappear, would have an immediate and catastrophic impact on society: nurses, garbage collectors, teachers, and dockworkers, for example. Our society would grind to a halt without people that care for others, teach others, make things, move things, and maintain things.

The transformation of work over the past century from manual labor to service sector work has been dramatic. Between 1910 and 2010, the number of people working in farming and industry and as domestic servants has shrunk. At the same time, the number of professional, managerial, clerical, sales, and service workers tripled, growing from one-quarter to three-quarters of total employment. Production jobs have been automated away. We have created new industries like telemarketing and financial services, while expanding corporate law, academic and health administration, human resources, lobbying, and public relations. Ancillary industries like dog washers and grocery delivery exist mainly because many people are spending so much time working in those other industries.

After the publication of Graeber's article in *Strike!*, a tax litigator contacted him to share, "I am a corporate lawyer. I contribute nothing to this world and am utterly miserable all the time." Telemarketers wrote Graeber to explain that they hate their jobs because they are paid to trick or pressure people into buying services they don't want or need.

IKIGAI AND THE ELEMENTS OF MEANINGFUL WORK

A Japanese term may explain why people in the Okinawa Prefecture have one of the longest life expectancies in the world. *Ikigai* means "reason for living," where *iki* means "life" and *gai* comes from the word *kai* ("shell" in Japanese) which was deemed highly valuable in the Heian period (794 to 1185). Older Okinawans can articulate the reason they get up in the morning, and living with purpose makes them feel needed well into the eleventh decade of their lives.

Viewing this term through the lens of American work culture, having *ikigai* at work means you are doing work that you're good at, the world needs, someone will pay you for, and that you love.

FIGURE 4: IKIGAI

Source: *Toronto Star*

Ideally, everyone who works would be experiencing *ikigai* at their job, and those who want to work should be able to find a job they're good at and that the world needs them to do.

Colorado State University professor Michael Steger studies meaningful work and explains the core components through his Work and Meaning Inventory. The three components include the degree to which people find their work to have significance and purpose, the contribution work makes to finding broader meaning in life, and the desire and means for one's work to make a positive contribution to the greater good.

Compiling their own findings with findings from other research teams, Steger found that meaningful work can consist of some or most of the following elements:

- skill variety
- opportunity to complete an entire task (task identity)
- a task's significance to other people
- military pride
- engagement
- a sense of calling
- challenge
- work role identity
- work centrality
- work values
- intrinsic work orientation
- spirituality
- good pay
- reputation[24]

Embedding these conditions into the next generation of startups, non-profits, and government agencies will ensure we meet human needs to take part in fulfilling work that the world needs done. The Great Pivot is about bridging the gap between where we are and where we want to be.

ALL HANDS ON DECK—TYPES OF JOBS NEEDED

A ship founded on the societal conventions of consumerism is sailing the U.S. straight toward the iceberg of accelerating negative global environmental trends. Those who understand global environmental trends—climate change, species mass extinction, and rapid conversion of natural materials into waste—have seen the existential threat and have called out to the captain in warning.

The problem is that our ship does not have just one captain to make the ultimate decision: the ship has 195 captains, one for each of the 195 countries the United Nations recognizes. Most of the captains understand the need to change the ship's trajectory and set sail for a more sustainable future, but a handful of people are telling the crew and passengers not to worry, that everything will be fine; there's no iceberg ahead, and we should stay the course. This is confusing to both the crew and the passengers.

If a few holdout leaders refuse to recognize the existential threat dead-ahead, should we the crew and passengers try to persuade them to change their minds, or should we start preparing for a different course without their permission? Those preparing lifeboats that can only hold a few people risk leaving the remaining crew and passengers on a sinking ship. Better to stay aboard and prepare to change course.

No less an august organization than the International Panel on Climate Change essentially raised the threat level for the climate to "all hands on deck" with the release of their 2018 report, which underscores the urgency of acting decisively to decarbonize economies in the next twelve years. Thousands of expert and government reviewers worldwide contributed to the document that bears its thesis as its title:

> "Global Warming of 1.5°C: An IPCC special report on the impacts of global warming of 1.5°C above pre-industrial levels and related global greenhouse gas emission pathways, in the context of strengthening the global response to the threat of climate change, sustainable development, and efforts to eradicate poverty."[25]

To accomplish dramatic reductions of greenhouse gas emissions, only a massive deployment of talent from different disciplines and all experience levels working together will do. We need people with expertise in all job families including:

- Accounting
- Administrative
- Construction
- Customer Service
- Engineering
- Facilities
- Finance
- Human Resources
- Information Technology
- Installations & Maintenance
- Legal
- Management
- Manufacturing & Production
- Marketing
- Operations
- Project Management
- Sales

Given how much work needs to be done to move humanity to a place where we live in balance with the environment that supports us, there is room for people with all kinds of different skills. To reduce carbon emissions and reduce waste in all sectors of our industrial and consumer economy, we need to deploy large numbers of people in the energy, transportation, agriculture, and manufacturing sectors, as well as work to restore healthy forests, waterways, and wildlife populations. So much work needs to be done.

ADDITIONAL RESOURCES FOR CHANGE MANAGEMENT

As we think through the resources needed to build a sustainable future, we should keep in mind what we are up against. The conventional way we generate energy, transport people and goods, handle materials, use food, and treat nature are not sustainable, and these systems are locked on autopilot. In addition, people are comfortable with routines. To overcome systemic and behavioral inertia to do things differently takes an extra level of effort. Thus, we need additional resources to facilitate change.

We need talented communicators who can tell us, through videos, photos, and animations, stories about what is possible. We need the most charming members of society with strong interpersonal and communication skills to cajole, coax, and persuade people into action. We need people to market greener products and services with lower environmental impacts. We also need people who can help reengineer and implement process and behavior changes that use less energy, water, and materials and generate less waste. Successful change management requires people with technical skills but also those with charisma and strong communication skills.

Imagine deploying a level of resources equal to the challenges humanity faces. Consider the sense of mission and purpose people would enjoy when working on teams to solve difficult problems. Think about a time when you were working on a project when everything seemed lined up to ensure your success: you and your fellow team members were working well together, and you had a shared goal and a shared sense of purpose.

This is what Angel Horne's job feels like. She loves her works as a Public Relations, Media, and Marketing Coordinator for the Lady Bird Johnson Wildflower Center in Austin, Texas. The Wildflower Center works to protect and preserve native plants and natural landscapes by inspiring the 185,000 annual on-site visitors and their webpage visitors to enjoy the beauty of native plants. Those who have travelled in Central Texas in the spring, when colorful fields of Bluebonnet, Indian Paintbrush, and Pink Evening Primrose wildflowers sit in full bloom along the highways, understand why the Wildflower Center is a cherished place to Texans.

Center staff not only plant and maintain beautiful native plant demonstration gardens, they also conduct seed collection and banking, rare plant monitoring and research, ecological restoration, propagate native plants, and offer botanical expertise. Horne's role is to work with environmental designers, invasive species specialists, field ecologists, horticulturists, and arborists to craft messaging about their work that will inspire the public. She explained, "I've been an interpretive guide in the field before, and during that work I was reaching groups of maybe six to sixty people at a time. Now, I'm trying to bring that experience and impact to broader audiences, and we don't always have the luxury of having everyone right here, where we can point to a specific butterfly or tree—we have to bring that experience to them through their screen or the pages of *Wildflower* (our magazine) and interpret the experience so that it leaves an impression."

Whether Horne is working with staff in guest experience, fundraising and membership, education, or many other areas of the organization, they all operate with a common purpose: to educate millions of people by showing them the beauty in native plants and thus inspire them to help restore natural landscapes.

CHAPTER HIGHLIGHTS

- The official unemployment rate, 3.7% in September 2018, did not include the 37 million people between the ages of 25 and 64 who were not in the labor force.
- Outsourcing, automation, the gig economy, and low levels of employee engagement contribute to the crisis in the world of work.
- Meaningful work has three components: the degree to which people find their work to have significance and purpose, the contribution work makes to finding broader meaning in life, and the means to make a positive contribution to the greater good.

2

PATHWAYS TO

THE AMERICAN DREAM

"The future belongs to those who believe
in the beauty of their dreams."
—Eleanor Roosevelt

In the early 19th century, Frenchman Alexis de Tocqueville toured America to study the political, economic, and religious character of the new American nation. Insights he gleaned over nine months of travelling, observing, and talking with people yielded an admiration for the emerging social state. In his book *On Democracy in America*, de Tocqueville reflected on the American Dream:

> "In democracies the love of physical gratification, the notion of bettering one's condition, the excitement of competition, the charm of anticipated success, are so many spurs to urge men onwards in the active professions they have embraced, without allowing them to deviate for an instant from the track."

De Tocqueville was enchanted by a people full of optimism about the possibilities awaiting them and the enthusiasm to pursue their dreams.

A century later, James Truslow Adams described the American Dream as the aspiration that "life should be better and richer and fuller for everyone, with opportunity for each according to ability or achievement regardless of social class or circumstances of birth." Adams believed that the path of upward mobility was available to all. This idea is shared among many in this nation full of immigrants and descendants of immigrants.

One of my friends, Duc Tran, immigrated to the U.S. from Vietnam when he was 15. In 1981, Tran and his mother (Tran's father died before he was born) sought to escape the political and economic chaos that consumed Vietnam after the war. His mother heard that the United Nations High Commission for Refugees had set up refugee camps in various nations in Southeast Asia, and Tran and his mother thought that if they could reach one of the camps by boat, they would have a good chance of eventually reaching a Western nation.

Tran and his mother set out in a crowded boat with several other people carrying an inadequate supply of food and water. On the open sea, a U.S. Navy ship picked up their boat, interviewed them, and delivered them to a refugee camp in Thailand where they were processed for asylum in the U.S. Tran, his wife, and his mother now live in California. They are employed and share a house they own. They are living the American Dream of setting goals and working toward them to create a meaningful, comfortable life for themselves in a politically and economically stable country.

MEANINGFUL JOB CREATION FOR THOSE STRUGGLING

Millions of others are not living the American Dream, though. Most Americans don't even have a personal financial safety net to fall back on should they suffer a household emergency. Research by the Pew Charitable Trust found that 60% of Americans lack $2,000 of savings to help them pay for an expensive financial shock like an illness or emergency car or house repair. They also found that "from 2014 to 2015, more than a third (34%) of U.S. households experienced fluctuations of annual income of 25% or more," further exacerbating financial insecurity.[26] While one emergency can knock most people off course, something bigger, such as a major illness, divorce, or job loss, can send others into bankruptcy.

The U.S. is a relatively wealthy country, yet many among our ranks are falling through the holes of our frayed safety net. A *New York Times* article from January 2018 entitled "The U.S. Can No Longer Hide from its Deep Poverty Problem" highlights a study by Oxford economist Robert Allen. By Allen's standard, the line of absolute poverty in rich countries sits at $4 a day, which means that 5.3 million Americans are absolutely poor by global standards.[27]

Rebuilding a trustworthy safety net for food, shelter, and healthcare ensures that, in the words of economist blogger Umair Haque, "failure does not become fatal." A number of the meaningful jobs outlined in the Great Pivot will provide Americans the income they require to meet their basic needs while also helping to build a circular economy.

Rescuing surplus prepared food is one example: at a time when one out of six Americans are food insecure, we can divert a large portion of the 40% of the food we throw away. Creating more jobs to redirect surplus prepared food and "ugly" produce is deeply meaningful work that must be scaled up. Matt Stepanovich, Lead Food Rescuer at RePlate in San Francisco, is one such person. He loves doing daily food rescues around the city, training other food rescuers, and maintaining client relationships.

In 2011, Stepanovich was in Japan with his family when the earthquake and tsunami hit. Upon witnessing the devastation, he sought out a role assisting in the emergency response effort. After meeting a community group providing meals to disaster victims who had lost their homes, Stepanovich volunteered to transport food to the site.

When he returned home to San Francisco, Stepanovich started driving for Uber but missed the meaningful volunteer work he had been doing in Japan. It only took one meeting with Maen Mahfoud, Executive Director of RePlate, to realize he had found his career.

Wouldn't it be great if we created more jobs for people who are struggling at the bottom of the socio-economic ladder, jobs that both help build a sustainable future and help people be self-sufficient? Examples abound of items we throw away but that still have useful life in them or that offer scrap value: food, consumer goods, clothing, and building materials. Reusing or recycling these instead of throwing them away can provide many jobs for people who have a tough time securing employment.

Ex-felons, so often stuck on the bottom rung of the ladder, have a particularly hard time. Figure 5 illustrates the breakdown of the 2.3 million people currently in jails or prisons, from which 626,000 people in the U.S. are released every year.[28]

FIGURE 5: PRISON POPULATION IN THE U.S.

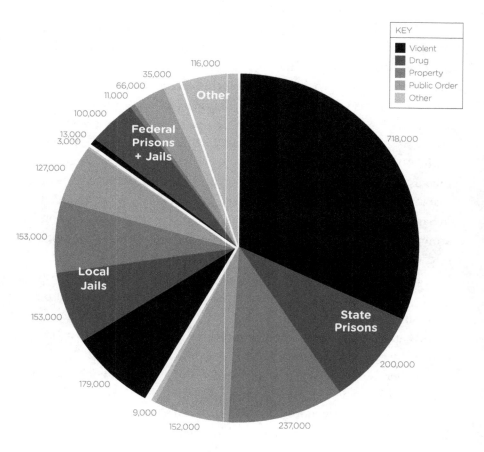

Source: Prison Policy Initiative

In order to reintegrate into society, people need jobs to support themselves. This is where social enterprises like HomeBoy Industries in Los Angeles come in.

Launched in 2011 by Kabira Stokes, the non-profit focuses on job training and electronic waste recycling for both gang-involved and previously incarcerated men and women. Stokes tells those about to be released from prison, "You get out and three days later come here, then start pulling apart a TV, and you have a job."

Over the past seven years, the company has employed 27 formerly incarcerated people who have recycled over 2.2 million pounds of electronics. The current CEO of HomeBoy Industries, Tom Vozzo, explained, "I was struck by the fact that the rest of the business world needs to understand that these folks who've faced barriers and overcome them are good, hard workers, and they should be employed."[29]

A private sector startup that's also working to employ ex-felons and people from vulnerable communities is New York City-based BlocPower. Their "Big Hairy Audacious Goal" is to retrofit a large portion of existing buildings in the U.S. for energy efficiency and to do so by hiring a large number of ex-offenders and people who grew up in public housing, have been in foster care, or have lived in low-income communities.

When BlocPower secures a contract with a municipality, utility, or building owner to install a more energy-efficient heating system or retrofit old lights for LEDs, they hire local construction companies. A condition in the contract stipulates that the construction companies must hire local folks from disadvantaged populations as part of their work crews. BlocPower has experienced initial push-back from construction companies who say they are not interested in running a social program, but that sentiment does not last long.

BlocPower works with top job-training programs in New York City and across the state to develop talent once they pass the qualification test. They believe the ex-felon *has* to be better at their job than another candidate who doesn't have a criminal record and thus rise to the challenge. CEO Donnel Baird explained:

"If you're giving somebody a shot who doesn't have a shot, they show up to work early, they work hard, they're really thankful and grateful for the opportunity, they want to make the most of it. They're just so much more motivated and passionate about the job opportunity they have."

Baird has noticed a shift in attitude from construction company managers after the formerly incarcerated start their jobs and prove themselves, saying, "You know what? I'm going to go out and figure out how to develop a pipeline of more of these workers, because they're willing to work twice as hard and twice as fast and twice as smart."[30]

Creating more meaningful jobs for ex-felons benefits society in another important way: by enabling them to support themselves and reintegrate into society, they will be less likely to reoffend and end up back in prison, thereby saving taxpayers tens of thousands of dollars per year, per prisoner.

For those who can find meaningful work, not just any job, such work can help them move up Maslow's hierarchy of needs. Figure 6 illustrates the higher-order needs that may be fulfilled when someone has a job that fills them with a sense of purpose.

FIGURE 6: MASLOW'S HIERARCHY OF NEEDS

Self-Actualization
Desire to Be the Most One Can Be

Esteem
Respect, Self-Esteem, Status, Recognition, Strength, Freedom

Love + Belonging
Friendship, Intimacy, Family, Sense of Connection

Safety Needs
Personal Security, Employment, Resources, Health, Property

Physiological Needs
Air, Water, Food, Shelter, Sleep, Clothing, Reproduction

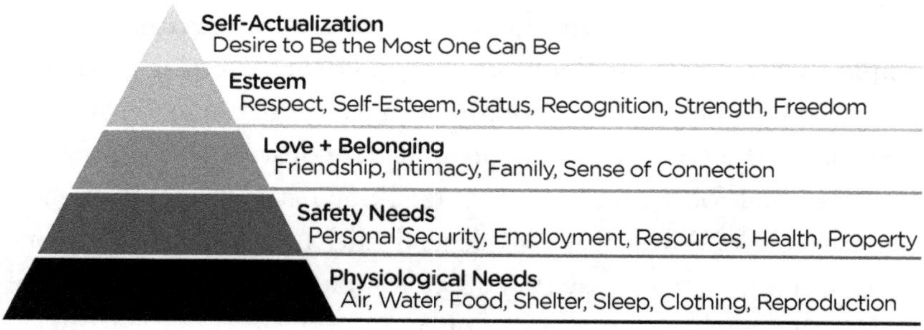

Source: psychologytoday.com

Given that work is an essential part of realizing the American Dream, perhaps part of the definition of the new American Dream is to have work that helps people meet their higher-order needs of self-esteem, respect, and self-actualization.

TRAINING AS THE ONRAMP TO MEANINGFUL WORK

The Building Materials Reuse Association (BMRA) conducts a three-day training for people who want to become certified deconstruction contractors. In 2017, the Oregon Department of Environmental Quality funded a three-day building deconstruction training for fourteen people in Portland.[31] Among those who attended the classes were people who were homeless, at-risk, or recently released from prison. The $3,500 cost per trainee was a small price to pay to help people develop valuable skills and become self-sufficient.

One trainee, a man named Jay who had recently completed a 25-year prison sentence, wrote in his application:

> "The only work experience I have is basic labor. No skills except the desire to move forward with my life and help provide for my family. I am strong, with a good work ethic. I believe in maintaining quality work performance and doing my job well."

Once he completed training and started work, his wife sent a touching thank-you letter to the training coordinator, expressing how much she appreciated the opportunity her husband had been given to rebuild his life.

After completing the BMRA classes, trainees continue to learn on the job. Scott Yelton at Portland's ReBuilding Center noted that on deconstruction jobs, workers experience the construction practice in reverse: they see how carpenters, plumbers, and electricians assembled the house in the past, thereby learning technical and construction skills. Yelton explained, "From my own personal experience, I've learned more about what NOT to do. For example, when taking out a window that has rotted, I've witnessed that the frame rotted because a header or flashing was not installed."

Yelton believes employees doing deconstruction also acquire soft skills equally valuable in the construction industry. They learn about communication, teamwork, time management, project management, and client services. On top of this, they are learning about health and safety best practices: how to safely use ladders, power tools, personal protective gear, and manage materials.

Jumping into a new line of work often requires training, either refreshing old skills or learning new ones. Training required to move into meaningful sustainability jobs may take a few days, a few months, or a few years. Some examples include:

- Leadership in Energy and Environmental Design (LEED) green building accreditation
- Project management certification
- Commercial driver's license
- Apprenticeship
- Licensing
- Business classes

Some of these trainings could be tailored to incubate new social enterprises. For example, Small Business Development Centers (SBDC), which have 900 delivery points around the U.S., could organize bundles of business classes on a regional basis with the express purpose of scaling up the number of people with workforce-ready skills. These workforce training programs could focus on nurturing some of the private sector projects listed in the Great Pivot in order to help those in the lower- and middle-income bracket.

MEANINGFUL WORK IS THE NEW AMERICAN DREAM

In the past few decades, the middle class has been destabilized by the growth of outsourcing, automation, and the gig economy. At the same time, the cost of the keys to the American Dream—college, childcare, and healthcare—has risen faster than inflation. Figure 7 shows the costs of goods and services that have risen and fallen most dramatically since the mid-nineties.

College tuition and textbook price increases far outpaced increases in costs of other consumer goods and services, with the result that 44 million Americans now collectively hold $1.5 trillion of student debt.[32] Rising childcare costs also place economic strains on households, as parents of small children depend on reliable childcare to be able to work. Then there's the cost of healthcare, which has outpaced inflation over the past twenty years as well. Healthcare spending in the U.S. in 2017

FIGURE 7: PRICE CHANGES OF SELECTED CONSUMER GOODS AND SERVICES (1996–2016)

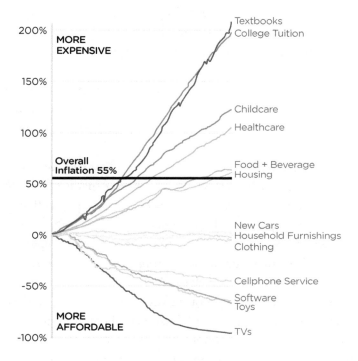

Source: Bureau of Labor Statistics

totaled $3.5 trillion for a population of about 325 million people—over $10,700 per person, by far the highest in the world.[33]

It is incumbent upon us as a society to help reduce the costs of the keys to the American Dream. At the same time, we must create meaningful jobs that pay enough for those struggling at the lower socioeconomic levels to be self-sufficient. We must also figure out the bare minimum everyone deserves in our wealthy society—food, shelter, healthcare, opportunities to work—as well as how society should cooperate for social benefit. The great flux in the world of work affords us a unique opportunity to redefine the social contract as well as the very nature of the American Dream.

Perhaps the American Dream is now stability, not upward mobility. We don't need to do better than our parents. If someone now graduates from college with debt in the low five figures, they're doing well. As long as we don't suffer a grave medical illness, the treatment for which requires us to sell our house or declare bankruptcy, we're OK. Winning the Powerball lottery jackpot or selling our startup to Google for an obscene amount of money would be nice, but at the end of the day, if we're free to pursue life, liberty, and happiness, we should feel like we've achieved the new American Dream.

Great wealth as an indicator of American success, after all, is inherently problematic. Economist-blogger Umair Haque urges Americans to strive for something more consequential than the material trappings of success.

> "Yesterday, we were taught that only in grandiosity lay safety—a shining career, a McMansion and so on. But that is not really the case; all these things, and the prices you must pay, will only leave you more prone to the anxieties of collapse, feeling more helpless and powerless to pay for them, maintain them, upgrade them . . . Yesterday's dream was a grandiose one, and now the question is whether we ourselves can develop a more grounded one. So how are we to give ourselves a sense of safety? Safety comes from a sense that one's life counts. Material things only create the illusion of counting. If your life is to count, then you must help turn this sense of anxiety into strength, power, grace, and courage."[34]

CHAPTER HIGHLIGHTS

- The middle class has been knocked off balance by work-place trends of outsourcing, automation, and the gig economy, as well as the fast-rising costs of college, child-care, and healthcare.
- Creating more meaningful jobs such as electronic waste recycling and building deconstruction and expanding training opportunities will help people at the bottom of the socioeconomic ladder become self-supporting.
- Meaningful work is one key to unlocking the new American Dream.

3

A VISION OF

A SUSTAINABLE FUTURE

"Now is not the time to shrink from the challenge of saving our only home in the universe. Now is not the time to pull into ourselves, retreating into either survivalist or escapist mode. To the contrary, this is the time for titans, not turtles. Now is the time to open our arms, expand our horizons, and dream big. Big problems require big solutions."
—Van Jones

When I worked at NASA's Ames Research Center as the sustainability manager in the Environmental Services Division, I hired professional storyteller Rafe Martin to lead a workshop about storytelling. My goal was to encourage the staff in my division to think through their role in building a sustainable future. Were we the stewards of the well-being of future generations, or were we the police of current environmental regulations? Or both? Being clear about our role influences which stories we tell, stories that in turn facilitate change.

My influence in bringing Martin to NASA was the CEO of Seventh Generation, Jeffrey Hollender. While searching for methods to make his company more sustainable, Hollender had enlisted Martin's help in training employees in the art of storytelling. Hollender wanted all 400 employees in his company, from mailroom staff to the Chief Marketing Officer, to look for opportunities for the company to reduce its environmental impact and to be able to verbalize these ideas effectively, both inside and outside the office walls. Similarly, I wanted my colleagues at NASA Ames to harness the power of storytelling to recruit others to help reduce the Center's environmental impact.

In service of this goal, one vivid story Martin told during the workshop at NASA showed the importance of having a vision. Martin explained that he loved riding motorcycles, and when he studied motorcycle safety, he learned the fine technical points of navigating a curve in the road. When a motorcyclist enters a curve, they need to keep their eye trained on the exit of the curve. At the same time, the rider subconsciously notes all the hazards: the gravel, the guardrail, and the car coming at them. They notice all these things in their peripheral vision, but they do not focus on them. If they take their eyes off the curve and look directly at any of the hazards, they risk a crash.

The current business as usual (BAU) trajectory of resource use and pollution generation will not lead us to a sustainable place where people live in balance with the environment that supports life. To create an alternate trajectory, we first need a destination in mind so we can travel down the right road.

The Brundtland Commission's definition of "sustainable development" from the 1970s provides a worthy destination: "Meeting our needs without sacrificing the ability of future generations to meet their needs." What we cannot seem to agree on are the technologies and systems that will drive us there. For the past few decades, we have debated the merits of nuclear power versus renewable energy, meat and dairy versus vegetarianism, wild-catch fishing versus aquaculture, and globalization versus regional economies. While we have not sketched out all of the details, we have clarity about many of the foundational parameters of sustainability and understand that the conventional way our economy uses resources and generates wastes is not helping us create a sustainable future.

Our current economy runs predominantly on non-renewable resources, uses materials in an inefficient and linear take-make-waste flow, and degrades Earth's natural systems. A sustainable future must include renewable energy, use resources efficiently and in a circular flow, and work to restore natural systems back to the point where they can regenerate themselves.

We need to continue having this conversation as a society about our vision for a sustainable future. Although it can at times be difficult and uncomfortable, approaching this conversation with an open mind and a commitment to finding solutions will help bring the currently inchoate vision into focus.

DOUGHNUT ECONOMICS

Part of the process of envisioning a sustainable future involves clarifying our goals. From a macro perspective, society should have two main sustainability goals: making sure every person's basic needs are met and staying within planetary limits. The core schematic from Kate Raworth's book *Doughnut Economics: Seven Ways to Think Like a 21st Century Economist* provides a helpful visualization of the floor and ceiling within which humans should exist (shown on the next page).

The inner ring explains the social foundation for humans, the bare minimum everyone needs: food, water, housing, energy, education, and networks. We also all deserve to have income and work, peace and justice, a political voice, social equity, and gender equality.

The outer ring of Raworth's graphic depicts the ecological ceiling below which human society must operate. Four particular areas require our immediate attention: we need to devote far more resources to rein in overshoot on climate change, biodiversity loss, land conversion, and nitrogen and phosphorus loading. By shoring up the shortfalls and reining in the overshoots, we will move to the safe zone of sustainability, which lies between the social foundation and the ecological ceiling.

As we will explore in this book, the waste of resources generated in one area, if allocated differently, could meet the needs of people suffering from insufficiency in another. Food waste is one example. In the U.S., we throw away 40% of the food we raise and grow while one in six Americans suffer from food insecurity. Diverting edible surplus food before it hits the waste stream and sending it to those who lack regular

FIGURE 8: DOUGHNUT ECONOMICS

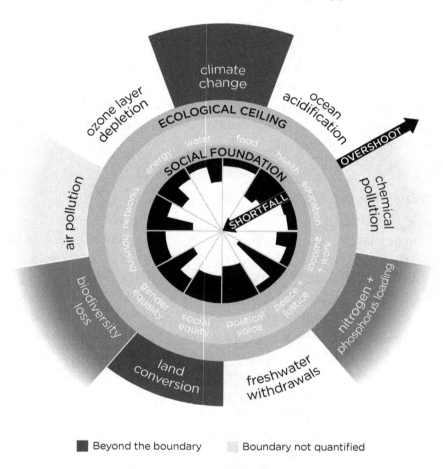

■ Beyond the boundary   ▨ Boundary not quantified

Source: Kate Raworth, *Doughnut Economics*

access to nutritious food will not only address hunger but will also reduce greenhouse gases, which contribute to climate change.

To avoid ecological overshoot, civilization must use resources no faster than the earth can regenerate them and generate waste no faster than the earth can absorb them, yet this is precisely what we are doing. One way to measure the degree to which different countries overshoot Earth's ability to provide resources and absorb waste is the Earth Overshoot Day, also known as Ecological Debt Day. The Global Footprint Network calculates the calendar date on which each country's resource

FIGURE 9: EARTH OVERSHOOT DAY, 2018

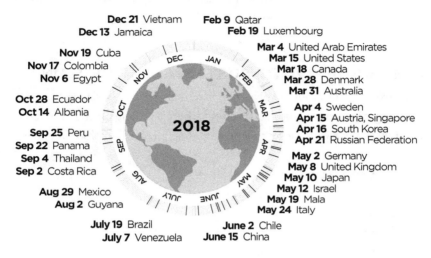

Source: Global Footprint Network

consumption for the year exceeds Earth's capacity to regenerate those resources. Figure 9 shows the overshoot day for dozens of countries.[35]

In 2018, the global average overshoot day fell on August 1. Countries that consume more resources overshoot sooner in the year, whereas countries that consume fewer resources overshoot later in the year. Comedian Johnny Steele tells a joke about overshoot in America, a country with one of the highest ecological footprints in the world: "If everyone on the planet lived like Americans, we would need five planets. Fortunately, Costco is going to start selling planets next year, and they're going to come in three-packs, so that'll be really convenient."

Satire news site *The Onion* also has something to say about living within our means. In the face of rising rents in Brooklyn, New York, *The Onion* announced: "Apartment Broker Recommends Brooklyn Residents Spend No More Than 150% of Income on Rent."[36] Obviously, at the household level, we would not take on basic monthly expenses beyond our monthly income, yet at the national level, we effectively do this. This happens in part because the effects of our transportation, housing, food, and lifestyle on natural systems are invisible to us. Earth does not have a Toyota Prius dashboard to inform us that when we stomp on the gas pedal, our gas mileage plummets. While we all intellectually understand we only have this one earth to provide our needs, most people do not have a sense of the scale of our environmental impact.

THE GREAT PIVOT'S SUSTAINABILITY VISION

The Great Pivot's vision involves five sustainability categories and provides a starting point for the discussion about where to devote limited resources. The five categories for job creation that this book will explore include:

1. Advanced energy communities
2. Low-carbon mobility systems
3. A circular economy
4. Reduced food waste
5. A healthy natural world

The first element of the Great Pivot's sustainability vision is advanced energy communities (AEC). By definition, AECs strive to meet zero net energy (ZNE) standards for the built environment by generating as much energy on-site as they use over the course of a year. They take full advantage of local renewable energy, demand response, solar emergency microgrids, and electric vehicle charging infrastructure. AECs add up to more than the sum of their parts by providing numerous co-benefits, including:

- Minimizing the need for new energy infrastructure
- Providing energy savings through ZNE
- Improving electric grid reliability and resilience
- Offering easier grid integration
- Providing affordable access to distributed energy resources and energy efficiency for all

Several projects detailed in the next chapter will help create meaningful jobs building advanced energy communities.

The second element in the Great Pivot's sustainability vision revolves around a low-carbon mobility system. Regional transportation systems should offer seamlessly integrated, low-carbon transportation options that help people move between regions, within regions, between towns, and around town. A low-carbon mobility system will be less dependent on single occupancy vehicles run by internal combustion engines and instead will invest in infrastructure that supports frequent and reliable trains, buses, shuttles, rideshare and electric bicycles, as well as first mile/last mile options of bicycles, electric scooters, and walking.

The third element in the Great Pivot's sustainability vision includes a circular economy. Material usage will be reduced by replacing single-use disposables with goods that can be reused at least 50 times, and the materials that remain will be recycled or composted regionally to create circular economy jobs.

Strategies to reduce food waste provide the fourth key piece of the Great Pivot's vision for a sustainable future. The percentage of food waste generated in the U.S. is vast, and rotting food makes such a large contribution to climate change that we cannot realize a sustainable future without tackling this issue. Food waste minimization strategies explored in this book include selling more ugly produce, diverting surplus prepared food to those who are food insecure, and helping consumers use more of what they purchase.

Finally, to make healthy our currently stressed and degraded natural systems, we must restore forests, topsoil, waterways, oceans, and wildlife populations to the point where they can regenerate themselves.

OVERARCHING VISION OF A STABLE CLIMATE

All five of these elements—building advanced energy communities, transitioning to low-carbon mobility systems, creating a circular economy, reducing food waste, and restoring natural systems—will help stabilize the climate, the major challenge of our time. To shift away from burning fossil fuels, which cause problems when they accumulate in our atmosphere, we need to implement technologies and systems that will both lower carbon emissions and sequester them. The blueprint already exists for how to bend down the business as usual curve.

Princeton professors Stephen Pacala and Robert Socolow developed the concept of climate stabilization wedges. Each wedge in Figure 10 represents a different strategy for bending down the BAU curve to a straight line. Stacked together, the wedges show what it takes to avoid the path of increasing greenhouse gas emissions that we're currently on.[37]

FIGURE 10: CLIMATE STABILIZATION WEDGES

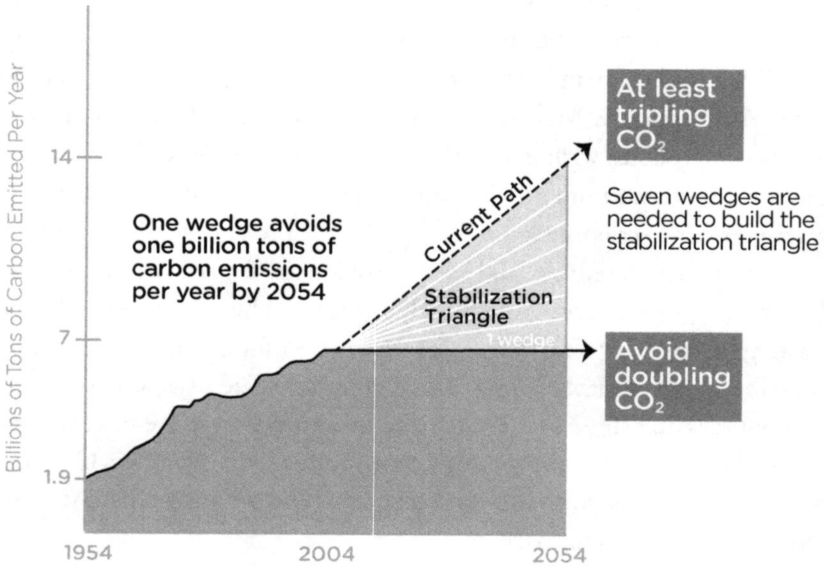

Source: Stephen Pacala and Robert Socolow

Pacala and Socolow's climate stabilization wedges fall into four sectors: electricity production, heating and direct fuel use, transportation, and biostorage. The level of effort needed to make dramatic changes in how our society generates energy seems like a tall order until we look around for inspiration.

STANFORD UNIVERSITY'S DRAMATIC GREENHOUSE GAS EMISSION REDUCTION

Stanford University, located in the San Francisco Bay Area, is attended by 16,500 students taught by 2,180 faculty and served by 12,150 staff spread out over 1,000 campus buildings on 8,180 acres. In 2007, Stanford University assessed its long-term energy needs and environmental sustainability goals and decided to do something special: the university developed and implemented plans for a Central Energy Facility (CEF) with a cutting-edge heat recovery system that takes advantage of the 70% overlap in campus heating and cooling needs. The CEF opened in 2015.

Located in a mild climate, Stanford heats and cools campus buildings simultaneously, year-round. Figure 11 shows the overlap. In winter there's more heating than cooling, in summer there's more cooling than heating, and the simultaneous heating and cooling provides the opportunity for improvement.[38]

FIGURE 11: 2016 STANFORD UNIVERSITY HEATING AND COOLING OVERLAP (HOT WATER & CHILLED WATER PRODUCTION SOURCE)

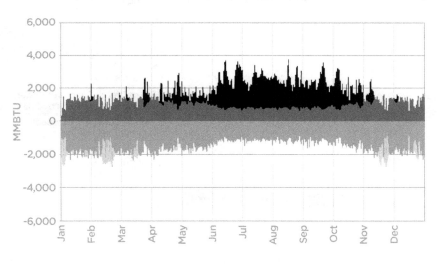

Source: Stanford University

The technologies utilized by the system are complex, but at the heart of it lies the concept that waste heat extracted from one area can be put to work in another area. The magic happens as heat is moved through a network of 22 miles of hot- and cold-water pipes between campus buildings and the CEF's hot and cold thermal-storage tanks. This heat exchange system, along with state-of-the-art demand management controls, allows for economical time-of-use operation so the school can buy electricity when utility energy rates are lowest, thus providing dramatic reductions of greenhouse gas emissions for the campus. Figure

12 indicates the huge drop in greenhouse gas emissions that occurred between 2014 and 2016.

FIGURE 12: STANFORD UNIVERSITY EMISSIONS REDUCTIONS WEDGE AND TARGETS

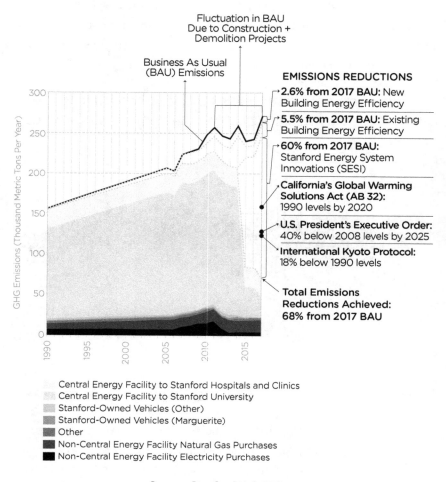

Source: Stanford University

The CEF system was expensive and does not have a quick return on investment (ROI), but there are other reasons the university designed and built the system. Even though the project had a long ROI, Stanford realized two major benefits from the project. First, by moving the systems that heat and cool the campus to another area, Stanford freed up valuable real estate in the center of campus for new academic buildings. Second, researchers and staff at the university created a state-of-the-art project that showcases innovative technologies and groundbreaking designs to inspire others to dramatically reduce greenhouse gases.

Now that Stanford has led the way and shown what is possible, waste heat recovery technologies like these should be replicated and scaled. The International Energy Agency estimates that unrecovered waste heat accounts for 47% of energy consumed globally. This could be a game-changer for the climate. Now we need to figure out how to invest in these changes on a larger scale.

Noblesse oblige refers to the responsibility of privileged people to act with generosity and nobility toward those less privileged. To bend the greenhouse gas BAU curve down, we need more environmental *noblesse oblige*. Those with more resources have the responsibility to develop, implement, and finance sustainability projects to bring down the costs for others and speed up the adoption of new sustainability projects. As Franklin D. Roosevelt said, "Great power involves great responsibility."

WE CAN'T AFFORD NOT TO DO IT

Some sustainability projects are too expensive for most of us, while others offer such attractive financials that there's no reason to keep wasting money on inefficient, outdated methods. Take coal-fired power plants, for example. In 2015, under the Obama Administration, the U.S. Environmental Protection Agency created the Clean Power Plan to identify and clean up the most polluting power plants, which were identified to be the older coal-fired ones with no air pollution emission controls. The Clean Power Plan originally aimed to cut carbon emissions 32%, to below 2005 levels, by 2030 and asked the states to determine the best ways to achieve the emission reductions.[39]

Since the Trump Administration took over, they have diligently worked to roll back the Clean Power Plan in spite of broad public support for cleaner air and addressing climate change. This is a losing battle for the Trump Administration, though: utilities cannot afford to keep inefficient, decades-old power plants in operation. A 2018 Sierra Club study[40] analyzed the cost of retiring 24 of PacifiCorp's coal-fired power plants, and their analysis found it would be cheaper for the utility to replace eleven with solar than to continue operating them. These coal units, located primarily in Utah and Wyoming, have a combined capacity of 2.73 gigawatts. From a business perspective, PacifiCorp can't afford not to replace coal with solar.

Replacing outdated technologies provide other benefits as well. When defenders of the status quo claim we cannot change because it will be too expensive, keep in mind that building something new will create jobs. That means more income for families and more money flowing through local economies.

While the Great Pivot's vision of a sustainable future involves advanced energy communities, low-carbon mobility systems, a circular economy, reduced food waste, and restored natural systems, the over-riding goal of all this work is mainly to stabilize the climate. It is the issue that supersedes all others. Without a stable climate, practically all other issues are moot.

BLUEPRINT FOR GREENHOUSE GAS DRAWDOWN

Paul Hawken's 2017 book *Drawdown: The Most Comprehensive Plan Ever Proposed to Reverse Global Warming* was tremendously exciting for those of us concerned about climate change. We have understood the science of climate change for years and have seen the ravages of drought, wildfire, and hurricanes, and have known we need to reduce the amount of fossil fuels burned, but we haven't been sure where to focus limited resources.

Paul Hawken and his team did the research and ran the modeling to draw up a blueprint. We now have a better idea about which projects will give us the biggest bang for the buck in drawing down atmospheric carbon over the next 30 years.

Some of the 80 projects detailed in *Drawdown* require huge budgets and large coordinating teams. Other projects can be implemented locally by small businesses. The solutions include projects that can happen all over the world, not just in the United States. (Some projects like Tropical Forests are clearly suitable only for tropical regions.)

Here is a summary of the 80 projects, ranked by the amount of carbon dioxide equivalent they will reduce if implemented globally.

TABLE 1: MOST IMPACTFUL SOLUTIONS TO DRAW DOWN
ATMOSPHERIC GREENHOUSE GASES

RANK	SOURCE	TOTAL ATMOSPHERIC CARBON DIOXIDE EQUIVALENT REDUCTION (GIGATONS)
1	Refrigerant Management	89.74
2	Wind Turbines (onshore)	84.60
3	Reduced Food Waste	70.53
4	Plant-Rich Diet	66.11
5	Tropical Forests	61.23
6	Educating Girls (in developing nations)	59.60
7	Family Planning	59.60
8	Solar Farms	36.90
9	Silvopasture	31.19
10	Rooftop Solar	24.60
11	Regenerative Agriculture	23.15
12	Temperate Forests	22.61
13	Peatlands	21.57
14	Tropical Staple Trees	20.19
15	Afforestation	18.06
16	Conservation Agriculture	17.35
17	Tree Intercropping	17.20
18	Geothermal	16.60
19	Managed Grazing	16.34
20	Nuclear	16.09
21	Clean Cookstoves	15.81
22	Wind Turbines (offshore)	14.10
23	Farmland Restoration	14.08
24	Improved Rice Cultivation	11.34
25	Concentrated Solar	10.90
26	Electric Vehicles	10.80
27	District Heating	9.38
28	Multistrata Agroforestry	9.28
29	Wave and Tidal	9.20
30	Methane Digesters (large)	8.40
31	Insulation	8.27
32	Ships	7.87

RANK	SOURCE	TOTAL ATMOSPHERIC CARBON DIOXIDE EQUIVALENT REDUCTION (GIGATONS)
33	LED Lighting (household)	7.81
34	Biomass	7.50
35	Bamboo	7.22
36	Alternative Cement	6.69
37	Mass Transit	6.57
38	Forest Protection	6.20
39	Indigenous Peoples' Land Management	6.19
40	Trucks	6.18
41	Solar Water	6.08
42	Heat Pumps	5.20
43	Airplanes	5.05
44	LED Lighting (commercial)	5.04
45	Building Automation	4.62
46	Water Savings (home)	4.61
47	Bioplastic	4.30
48	In-Stream Hydro	4.00
49	Cars	4.00
50	Cogeneration	3.97
51	Perennial Biomass	3.33
52	Coastal Wetlands	3.19
53	System of Rice Intensification	3.13
54	Walkable Cities	2.92
55	Household Recycling	2.77
56	Industrial Recycling	2.77
57	Smart Thermostats	2.62
58	Landfill Methane	2.50
59	Bike Infrastructure	2.31
60	Composting	2.28
61	Smart Glass	2.19
62	Women Smallholders	2.06
63	Telepresence	1.99
64	Methane Digesters (small)	1.90
65	Nutrient Management	1.81
66	High-Speed Rail	1.52

RANK	SOURCE	TOTAL ATMOSPHERIC CARBON DIOXIDE EQUIVALENT REDUCTION (GIGATONS)
67	Farmland Irrigation	1.33
68	Waste-to-Energy	1.10
69	Electric Bikes	0.96
70	Recycled Paper	0.90
71	Water Distribution	0.87
72	Biochar	0.81
73	Green Roofs	0.77
74	Trains	0.52
75	Ridesharing	0.32
76	Micro Wind	0.20
77	Energy Storage (Distributed)	N/A
77	Energy Storage (Utilities)	N/A
77	Grid Flexibility	N/A
78	Microgrids	N/A
79	Net Zero Buildings	N/A
80	Retrofitting	N/A

Source: *Drawdown*

These projects all exist and are scaling to become competitive alternatives to the carbon-intensive technologies that currently dominate our world. More details about the technologies, costs, and savings of these projects can be found at drawdown.org.

THE 30 GREAT PIVOT PROJECTS

Drawdown explained which measures will provide the greatest potential for drawing down greenhouse gases in the atmosphere. The Great Pivot then picks up where *Drawdown* leaves off by detailing 30 projects, with their proofs of concept, that can create millions of jobs in the U.S. In chapters four through eight, we will take an in-depth look at projects in energy, transportation, the circular economy, food waste, and restoring the natural world with an eye toward better understanding how we can replicate and scale the following projects.

(1) Zero Net Energy Retrofits—Single-Family Homes

(2) Zero Net Energy Retrofits—Multi-Family Dwellings

(3) Solar Emergency Microgrid Retrofits at Hospitals and Municipal Emergency Response Centers

(4) Commercial Solar Siting Surveys for Counties

(5) Zero Net Energy Retrofits—Commercial

(6) Part-Time Embedded Sustainability Project Manager

(7) Development of Mass Transit

(8) Transportation Management Associations

(9) Development of Safe Bicycling Infrastructure

(10) Mobility-as-a-Service Apps

(11) Electric Vehicle Charging Infrastructure

(12) Designing Walkable Communities

(13) Local Government Waste Prevention Coordinators

(14) Building Deconstruction

(15) Tool Lending Library + Repair Cafe + Maker Space

(16) Upcycling Dead or Diseased Trees

(17) Artistic Upcycling and Salvage

(18) Regional Recycling Market Development Managers

(19) Reverse Catering

(20) Community Kitchens

(21) Ugly Produce Distribution

(22) Business Services Supporting Small, Organic Farms

(23) Carbon Farming

(24) Restoring Healthy Forests

(25) Construction Products and Furniture

(26) Restoring Healthy Waterways

(27) Wildlife Defense

(28) Wildlife Restoration

(29) Wildlife Overpasses and Underpasses

(30) Rewilding for Habitat Restoration

Keep in mind that this list of 30 projects is not comprehensive but, rather, a proposal for next steps. The hope of the Great Pivot is to catalyze a community of people who want to help build a sustainable future, both those in a position to implement policy changes and those who want to secure meaningful, fulfilling employment. By being more proactive about creating not just renewable energy jobs—which are important— but many other types of sustainability jobs, we will move humanity to the safe space between the social foundation and the ecological ceiling and secure the health of the planet.

CARING FOR THE BIG BLUE MARBLE

For several decades now, we have seen images of our precious Earth from space. In 1946, the first grainy and barely recognizable images were sent from a V-2 rocket. In 1947, photos were taken from 100 miles above Earth. We have seen the "Earthrise" photo from the moon, which was taken in 1968. Then, in 1972, we saw the stunning Big Blue Marble image. That was the year Apollo 17 sent back numerous images of Earth, which were assembled into a composite picture. The image of the whole planet, in which Africa prominently figures, offers a spectacular view of our precious home.[41]

Astronauts who have visited the moon or the International Space Station (ISS) have commented on how beautiful and delicate our planet looks from space. In 2016, astronaut Scott Kelly described the haze over Asia and Central America and said, "When you look at the atmosphere on the limb of the Earth, I wouldn't say it looks unhealthy, but it definitely looks very, very fragile and just kind of like this thin film, so it looks like something that we definitely need to take care of."

NASA's work to study our home planet and explore beyond yields bountiful wisdom about how we should live more sustainably. One of my former colleagues at NASA Ames, researcher Dr. John Hogan, works in Advanced Life Support systems. While studying advanced life support systems needed to keep humans alive in the ISS, Dr. Hogan realized that the ISS provides insights into what we do and do not know about Earth's life support systems.

He explained that on the ISS, astronauts need carefully maintained air, water, and material life support systems to stay alive. Systems vigilantly balance the components of the air to keep each of the gases at the exact threshold needed to support life, and continuously treat and recirculate the water the astronauts use. For trash, astronauts basically scrunch it up, wrap it in duct tape, and store it until the Soyuz spacecraft arrives with deliveries of fresh supplies and take the waste back to Earth. The ISS is not a closed-loop system, since it regularly receives new supplies and ships out waste. As NASA plans to send astronauts to Mars for extended periods of time, they know they need to engineer truly closed-loop systems that receive no new inputs.

As NASA researches those closed-loop life support systems, the lessons learned teach us what we do and do not know about life support systems on Earth. Dr. Hogan explained that with the ISS, we know the precise thresholds for each advanced life support system above or below which life cannot be supported. We also know that on the ISS, advanced life support systems have small buffers, whereas on Earth the buffers are much larger. Oceans dilute pollution. Wind blows pollution "away." While the thresholds we must not cross in space are clear, the thresholds for Earth's life support systems are less so.

We don't know how much degradation the carbon cycle, nitrogen cycle, forest ecosystems, marine ecosystems, and the food web can handle before they will be unable to bounce back. Because we do not know enough about Earth's life support systems, to hedge our bets we should ramp up restoration efforts. A key part of the vision of a sustainable future must involve healthy, vibrant natural systems.

CHAPTER HIGHLIGHTS

- *Doughnut Economics* posits two main sustainability goals: making sure every person's basic needs are met and staying within planetary limits. Humans should exist within the safe space between the social foundation and the ecological ceiling.

- The Great Pivot's sustainability vision includes five categories: advanced energy communities, low-carbon mobility systems, a circular economy, reduced food waste, and a healthy natural world.

- Dramatic reductions of greenhouse gases are possible. Stanford University's GHG emissions fell 68% mainly by investing in a new campus-wide heating and cooling system.

4

MEANINGFUL JOBS FOR

ADVANCED ENERGY COMMUNITIES

"Organisms sip energy because they have to
work or barter for every single bit that they get."
—Janine Benyus, author of *Biomimicry*

After purchasing a 110-year-old folk-Victorian farmhouse in Ann Arbor, Michigan, Matt Grocoff and his wife started renovating it in 2006 for maximum energy efficiency and to generate as much on-site energy as it uses, that is, ZNE. In a climate with a challenging 100-degree temperature swing between 95° humid heat in the summer and sub-zero temperatures in the winter, they knew that an insulated building envelope was vital.

The last week of January 2019, the polar vortex hit, sending temperatures down to –35° with the wind chill. Ford and General Motors shut down production. The local utility company issued an emergency alert to cell phones asking people to turn down their furnace to 65° because a fire at a compressor station threatened the entire gas grid. Grocoff's house, however, doesn't use gas because it's ZNE.

During the cold snap, Matt's four-year-old-daughter said, "I love our house." When he asked why, she said, "Because it's toasty warm."

Grocoff explained what he learned during his ZNE retrofit: to be successful at achieving ZNE, the most important thing is commitment to the goal. Once the target is set, a ZNE retrofit basically involves three main steps. "The first is to electrify everything. Switch your water heater, furnace, and clothes dryer to electric heat pump and stove to electric induction. By nature, electric appliances are more efficient, more durable, require less maintenance, and are far safer than gas." Electrification is a precondition to meeting global climate goals. You can't power a gas furnace with renewable energy. To eliminate fossil fuels, we must eliminate the source of the demand.

The second step is to blanket the home with as much insulation as you can and air seal the house as tightly as is feasible by either installing high-efficiency windows and doors or weather-sealing the existing ones (Grocoff opted to restore his home's original windows).

The final step is to install on-site renewable energy to generate electricity. All the energy the Grocoffs' house needs is produced by an 8.1kW solar photovoltaics array on the roof and a ground-source heat pump (geothermal). Ground-source heat pumps run air or water through a series of subterranean loops to tap the near-constant 57° heat deep underground, then concentrate the heat in the air or water and put the heat to work aboveground.

Electric appliances powered by on-site renewable energy avoid two problems that contribute to climate change: inefficient waste heat that occurs when burning natural gas on site, and the electricity transmission and distribution losses that occur when electricity is generated at centralized power plants and transmitted great distances.

The predominant ways the U.S. generates and uses energy is not sustainable, and there is much work to be done in this sector. In 2017, according to the U.S. Energy Information Agency, 32% of U.S. electricity came from natural gas, 30% from coal, 20% from nuclear, and 17% from renewables.[43] Our current electricity grid is not only dirty, with a majority of U.S. electricity generated by the burning of fossil fuels, it's also vulnerable: one tree falling on a power line can take down power for millions of people. We deserve a cleaner, more resilient, and more reliable electric grid.

The goal of the Clean Coalition, a California nonprofit, should be the goal of our entire nation: "To accelerate America's transition away from an outdated energy system built around large, centralized, fossil fuel power plants and miles of inefficient transmission lines toward a modern energy system where smaller-scale, efficient, renewable energy projects deliver affordable and reliable power to communities." A key building block for this clean, resilient-energy future should be advanced energy communities. The Clean Coalition describes AECs as having the following components:

- Abundant solar electricity, energy storage, and other distributed energy resources
- Low or zero net energy buildings
- Solar emergency microgrids for power management and support of critical facilities during outages
- Charging infrastructure to support the rapid growth of electric vehicle (EV) usage

With the goal of scaling up this type of clean energy infrastructure, let's look at projects that will create meaningful jobs and help build an AEC future.

PIVOT #1—ZERO NET ENERGY RETROFITS— RESIDENTIAL BUILDINGS

The first of the 30 meaningful job creation pivots involves retrofitting single-family homes for ZNE. The vast majority of the millions of homes in the U.S. have leaky single-paned windows, uninsulated walls, inefficient lighting without motion sensors, uninsulated water heaters, and inefficient furnaces that burn heating oil or natural gas. Only a small fraction of U.S. homes have on-site renewable energy such as solar panels or ground-source heat pumps. Much work needs to be done to retrofit millions of homes in the U.S. for ZNE.

To become ZNE would mean: making the building envelope and electrical and mechanical systems as efficient as technologically possible; replacing technologies that burn fossil fuels on-site with technologies that use electricity (called electrification); then installing enough on-site renewable energy so that the building generates as much electricity as it uses over the course of a year.

A ZNE retrofit involves multiple steps, each with a steep learning curve. My brilliant husband, who cares deeply about climate change and has a full-time job, has been working diligently for years to make our home ZNE. He's still not done. Looking at the steps involved gives a sense of why retrofitting for ZNE can take homeowners so long. The process includes:

- Researching and hiring a firm to do an energy-efficiency audit
- Researching the cost and quality of energy-efficient technologies
- Researching energy-efficiency retrofit contractors
- Researching and securing rebates and financing for energy-efficiency technologies
- Researching the cost and ratings of solar contractors
- Researching and securing tax incentives and financing for solar panels

Professionals are available to help with each of these different tasks, but it takes time to find them all. Utilities, municipalities, contractor firms, and financial institutions can help with various pieces, but for the most part, they're not working together. One solution to accelerate the pace of ZNE retrofits is a one-stop shop that can oversee the project management for the homeowner, then roll all costs and savings into the utility bill. Until that happens, motivated individuals will continue to take it upon themselves to coordinate contractors and financing for their home's ZNE retrofit, but most people will not. (There's more on this topic below.)

Having more stringent state building energy efficiency codes and stronger incentives from energy utilities would help as well. Otherwise, the steep learning curve people endure to accomplish something complicated like a ZNE retrofit will ensure only a small fraction of people will do it.

Regulatory Framework Needed for Progress

Stronger building codes make a difference, and California offers a case in point. In the 1970s, California established its Title 24 building code, which progressively tightened energy efficiency requirements in three-year cycles. Over time, this has resulted in a nearly flat per capita energy use compared to the rising rate throughout the rest of the U.S. This effect, shown in Figure 13, is called the Rosenfeld Effect, for Art Rosenfeld who championed energy efficiency in California.

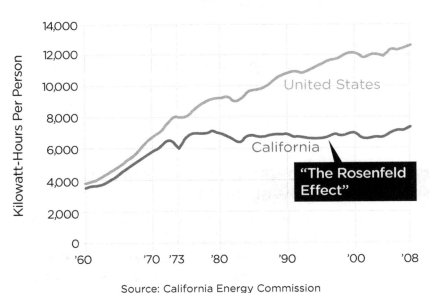

FIGURE 13: ROSENFELD EFFECT

Source: California Energy Commission

Not content to stop there, starting in 2020, California will require that all new residential construction be ZNE. This requirement applies to new homes, not existing ones. For existing buildings, California's Title 24 three-year regulatory cycle will continue to tighten up energy-efficiency requirements and require more solar systems on existing buildings to drive ZNE upgrades.

Existing building stock needs to be upgraded. If we look at the number of existing single-family homes in just four states, we see how much work lies ahead of us.

- California: 9 million single-family homes
- Michigan: 3.1 million single-family homes
- Louisiana: 1.3 million single-family homes
- Virginia: 2.1 million single-family homes

Think of the number of jobs we could create for teams of people to do ZNE retrofits throughout the country.

Jobs needed:

- Project managers
- Energy efficiency auditors
- General contractors
- Electricians
- Heating Ventilation and Air Conditioning (HVAC) specialists
- Solar installers
- Finance specialists

PIVOT #2—ZERO NET ENERGY RETROFITS—MULTI-FAMILY DWELLINGS

Persisting through the long, arduous process of implementing ZNE retrofits for a single-family home takes dedication. Doing this for multi-family dwellings (MFD) poses even more of a challenge due to additional layers of complexity. One such hurdle comes from a "split incentive," which blocks progress on ZNE retrofits by building owners and tenants.

The split incentive problem pertains to rental properties. In MFDs, the landlord owns the building but usually does not pay the energy bill. This means there is not a direct incentive for the property owner to invest in energy-efficiency upgrades and on-site renewable energy, since the owner will not realize energy bill savings. At the same time, tenants pay the energy bill and would save money if energy efficiency and solar were implemented, but they are hesitant to invest in improving a building they do not own.

Utilities can help overcome this impasse by financing the retrofits, then rolling the costs and savings into the energy bill. The main benefit to doing this for utilities in areas with growing populations is that aggressively implementing energy efficiency will allow the region to avoid the enormous expense of building new power plants. In areas where populations are not necessarily growing, utilities that want to help their low-income customers will invest in their well-being by helping lower their monthly utility bills.

Pay As You Save in Southern Arkansas

Ouachita Electric Cooperative in southern Arkansas implemented an impressive Pay As You Save (PAYS) program for their low-income customers who live in MFDs. Ouachita General Manager Mark Cayce explained that many of their 9,400 customers are low-income or seniors, some of whom were paying $400 or $500 per month for their energy bills in the hot summer months.

Seeking to reduce utility bills for their customers, Ouachita Cooperative invested $6,500 per unit in energy efficiency. The process started when the cooperative approached MFD building owners and asked if they could send certified energy efficiency auditors to identify opportunities and install them at no cost to the building owners. "I haven't found a landlord yet who would turn us down," Cayce explained. When Ouachita asked the building tenants if the cooperative could install heat pumps, weather stripping, insulation, Nest thermostats, and LED lights for free and lower their energy bills by at least 20%, all tenants said yes.

The main reason Ouachita Electric Cooperative did the retrofits free of charge was to help their customers rein in peak demand prices. In the hottest summer months, when temperatures regularly top 90° and 100°, air conditioning in uninsulated buildings drives up bills and pushes up peak pricing. In fact, energy prices for the whole year are set by peak summer demand. Shaving down peak demand moderates energy prices for ratepayers the rest of the year.

Ouachita Electric Cooperative in southern Arkansas has retrofitted 85 MFD units to date with funding from the energy efficiency tariff (a separate line item on each cooperative customer's energy bill) and financing from the Rural Utility Service, a branch of the U.S. Department of Agriculture. With 0% financing over twenty years from the Rural

Utility Service, Ouachita plans to help more customers make energy efficiency improvements and cut their energy bills.

Offering MFD energy customers solar photovoltaics is the next project. Cayce plans to install 100 kilowatts of solar on an MFD complex in 2019 at a cost of $200,000. This project makes sense in the area, as peak energy demand matches the time when solar panels generate the most electricity.

Replicating Pay As You Save Programs

Slowly, Pay As You Save (PAYS) programs like these are spreading throughout the country. Utility companies are realizing that PAYS offers an irresistible tool for overcoming the barriers to energy efficiency upgrades in MFDs. One impetus for PAYS program implementation around the country has been Vermont-based non-profit Energy Efficiency Institute.

Harlan Lachman, President of Energy Efficiency Institute, has helped utilities in Arkansas, Kansas, New Hampshire, and Hawaii set up PAYS programs. Based on his experience, Lachman recommends the following process for other utilities that want to create their own PAYS programs.

1. **Find a champion**—Having a high-ranking, local champion in the utility or the public utility commission is key. This person will advocate for the formation of a PAYS program and provide the leadership to get the program started.
2. **Find the capital**—The champion will also find funding to pay for energy efficiency upgrades at MFDs.
3. **Create a list of contractors**—Collecting and vetting a list of registered contractors approved to perform energy efficiency upgrades, as some utilities do, will accelerate implementation.

Lachman explained that in his experience, utilities that have decoupled their energy sales from profits (meaning their regulating authority structures incentives so that their profits are not tied to how much energy they sell) are the ones that have the most success setting up PAYS programs.

For those who want to go beyond the standard PAYS program of

MFD energy efficiency retrofits, utilities can also offer solar photovoltaic installations and roll the costs into the utility bill. Doing so would help reduce pressure to build new power plants, stabilize the climate, and create meaningful jobs.

Millions of MFDs throughout the country could be upgraded with ZNE retrofits, including:

- California: 3.1 million multi-family dwellings
- Michigan: 1.8 million multi-family dwellings
- Louisiana: 600,000 multi-family dwellings
- Virginia: 820,000 multi-family dwellings

Jobs needed:

- PAYS program managers at utilities
- Contractors to do energy efficiency and solar installations

PIVOT #3—SOLAR EMERGENCY MICROGRID RETROFITS AT HOSPITALS AND MUNICIPAL EMERGENCY RESPONSE CENTERS

Another type of ZNE retrofit focuses on hospitals and municipal emergency response centers. Retrofitting these two types of buildings is not just the right thing to do for the environment but will also allow emergency responders to better serve their communities in the event of a natural disaster.

The spate of natural disasters around the country in 2017 and 2018 highlighted the importance of local communities having seamless, functioning emergency response services. Wildfires in the Western U.S. and hurricanes in the Gulf Coast and Caribbean knocked out power for extended periods of time. When this happens on the mainland, hospitals and emergency response centers fire up diesel generators to provide continuous power. The upside of diesel generators is that they come online quickly; the downsides are that they burn fossil fuels, release air pollution, create noise pollution, are expensive to maintain (requiring monthly testing), and sit idle most of the time.

When disaster strikes and the electric grid goes down outside the mainland, as it did in Puerto Rico after Hurricane Maria in 2017, many areas go without power for extended periods of time. Immediately after Hurricane Maria, the National Guard—on behalf of the Federal Emergency Management Agency (FEMA)—delivered diesel generators to the most critical operations around the island until power could be restored. Periodically, the generators ran out of diesel fuel and the critical operations were again without power until resupplies arrived. Some remote medical clinics did not receive diesel generators at all and went without power until the grid was restored months later.

For critical operations on and off the mainland U.S., a cleaner and more resilient option for power would be solar emergency microgrids. This system combines solar photovoltaic panels, energy storage, and monitoring, communications, and controls to deliver electricity instantly and indefinitely for disaster response. A solar emergency microgrid is linked to the main electric grid but during a power outage can isolate itself and keep running, providing backup power for critical loads at priority facilities such as hospitals, police and fire stations, emergency operations centers and shelters, and critical communications and water infrastructure. This means that during normal operations, they would have clean power and, in emergency situations, they would have indefinite backup power, which would allow them to better respond to their communities' needs.

Setting up solar emergency microgrids for municipal emergency response centers, hospitals, and medical clinics in areas of the U.S. most vulnerable to natural disasters will involve a good chunk of work. Given the trend of an increasing number of natural disasters, our country would be well served by investments in solar emergency microgrids.

Consider just the number of hospitals and municipalities in these four states that would benefit from Solar emergency microgrid retrofits:

- California: 345 hospitals, 482 municipalities
- Michigan: 168 hospitals, 533 municipalities
- Louisiana: 233 hospitals, 308 municipalities
- Virginia: 166 hospitals, 95 counties, along with 38 independent cities

Jobs needed:

- Developers
- Financiers
- Engineers
- Project managers
- General contractors
- Electricians
- Solar system installers

PIVOT #4—COMMERCIAL SOLAR SITING SURVEYS FOR COUNTIES

When we think of green jobs, solar panel installation often pops to mind. The growth of solar photovoltaics installations over the past decade has created many meaningful jobs, the majority in utility solar projects. While growth has accelerated, we have only just started tapping the potential for solar energy in the U.S. To further speed up the build-out of solar, particularly in the commercial solar area, we need to map sites well suited for future installations.

The Solar Energy Industries Association tracks the growth of solar installations. Figure 14 shows the growth of solar from 2000 to 2016, with utility solar eclipsing the number of residential and non-residential (commercial) projects.

The Solar Foundation tracks the number of jobs created each year in the solar industry. Their 2017 National Solar Jobs Census calculated that the solar workforce increased by 168% in the past seven years, from about 93,000 jobs in 2010 to over 250,000 jobs in 2017.[44]

For a variety of permitting and grid interconnection reasons, which we will not go into here, commercial solar lags utility and residential solar. But one thing we can do to help commercial solar installations catch up to the number of utility and residential solar installations is map where commercial installations in urban and suburban areas could go. One model for how to do this comes from the Clean Coalition.

In San Mateo County in northern California, the Clean Coalition used Google Maps to examine an area with dense development and an

FIGURE 14: ANNUAL U.S. SOLAR PHOTOVOLTAICS INSTALLATIONS, 2000–2016

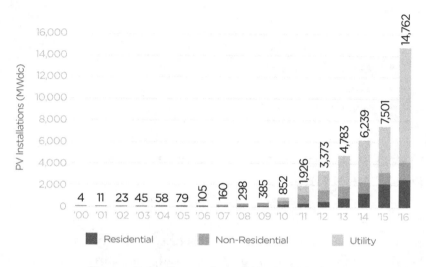

Source: Solar Energy Industries Association

extensive tree canopy to find possible locations for systems of at least 100kW of commercial-scale solar. They found sites in three main areas: K-12 school rooftops, parking lots, and parking garages. Then they assessed the sites for ease of interconnecting with the electricity grid to determine the cost-effectiveness of the solar projects. Adding up the cost-effective solar potential in the southern part of the county, the Clean Coalition's "Solar Siting Survey" identified 65MW of commercial-scale solar potential.

Every area of the country with strong solar potential and dense development should undertake this exercise. The best way to start is by looking at the U.S. Department of Energy's National Renewable Energy Lab map of solar potential in Figure 15.

The darkest shading on the map indicates the places offering the highest solar potential, where we should first deploy resources to build commercial solar on rooftops, parking lots, and parking garages. Creating more solar siting surveys, starting with the darkest parts of the map, would create projects to accelerate future build-outs.

FIGURE 15: SOLAR POTENTIAL IN U.S.

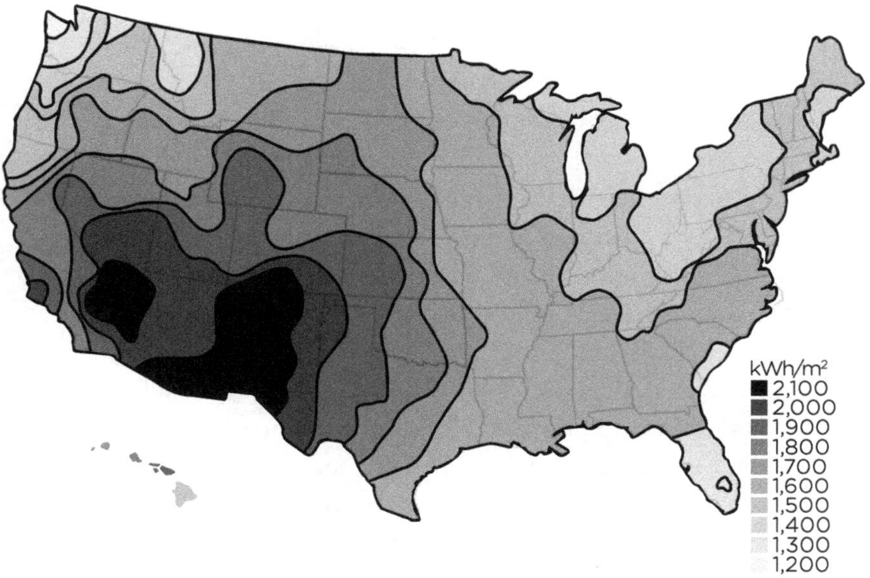

kWh/m²
- 2,100
- 2,000
- 1,900
- 1,800
- 1,700
- 1,600
- 1,500
- 1,400
- 1,300
- 1,200

Source: National Renewable Energy Lab

Jobs needed:

- Engineers
- Project managers

PIVOT #5—ZERO NET ENERGY RETROFITS—COMMERCIAL

In Sunnyvale, California, the building at 435 Indio Way was previously a boring, one-story, concrete tilt-up commonly found in 1970s Silicon Valley. Developer Kevin Bates purchased the building with the interior ripped out to the studs and retrofitted it into a healthy, delightful ZNE space. The building now features:

- Solar photovoltaics
- Electrochromic glass

- Daylight harvesting
- Operable skylights and windows
- Ceiling fans
- R-20 wall insulation
- R-40 roof insulation
- An efficient, 80% smaller HVAC unit

A few of these elements provided ancillary benefits. Electrochromic glass has higher insulating properties and reduces glare, which in Bates's building eliminated the need for blinds. Operable skylights and windows deliver natural air exchange and night-purge ventilation to freshen up indoor air. R-20 wall insulation on the outside of the building instead of the inside created an additional 326 square feet of leasable space. A smaller HVAC system meant smaller reserve requirements for replacement equipment and lower quarterly maintenance costs.

Finances for this project are dazzlingly attractive as well. Bates explained that the building is worth $100.29 more per square foot than if they had renovated it to meet minimum building code requirements. He spent $49 per square foot above and beyond a normal commercial space renovation. The additional value came from:

- $83.08 per square foot in reduced operating costs over 15 years
- $36.92 per square foot in above-market rent over 15 years
- $22.81 per square foot additional rent due to early leasing (Bates rented out the space within three months instead of the usual average of eighteen months that commercial properties sit vacant)
- $7.32 per square foot additional leasable space of 326 square feet

When studying before-and-after pictures of the building, the difference in natural light is striking. Before the renovations, the vast, empty space appears dark and cavernous. After the retrofit, with diffusing skylights, lightly colored concrete flooring, and white fabric on the ceiling to bounce light into the interior of the space, we can see how the space needs no artificial light during the day. Tenants love the space, as it was designed to enable 24/7 connectivity with the outdoors and utilizes nighttime air flushing and natural light. Bates tries new combinations of green building technologies on every ZNE retrofit project to see how much more pleasant, healthy, and energy efficient he can make the commercial space for tenants.

He explained, "You can do what a typical developer does and follow code, or you can jump in front of that wave and ride it. The wave is coming anyway. The benefit [of doing ZNE retrofits] is that your property will be more resilient in a down market and will allow you to differentiate yourself. The bottom line is that you're going to stay leased when everyone else is having trouble. It's not rocket science. By having less-complex systems with fewer moving parts to break down, like skylights, a thermal mass, and air exchange, a simpler building will be easier to take care of. It's lower risk and more profit."

He insisted that this is the new Class A office space. Developers who don't get on board risk being left behind.

Considering that commercial buildings turn over tenants every five to seven years,[45] it is easier for building owners to make improvements like energy efficiency upgrades and solar installations when the space is vacant. Retrofits can also be done while leased space is occupied though.

Sonic's ZNE Retrofit

Mynt Systems in Santa Cruz, California, performs ZNE retrofits for office buildings, light manufacturing facilities, schools, and museums. In 2017, Mynt retrofitted Sonic, an 82,000-square-foot electronics manufacturing facility in Fremont, California, for ZNE while employees occupied the building. Mynt scheduled construction during off hours and from time to time asked management to move employees to other parts of the building if their work allowed.

Upgrades included a new HVAC system, LED lights, diffusing skylights, solar photovoltaics, and electric vehicle charging stations. In combination, these technologies resulted in a ZNE building with more natural light and healthier air.

The project upgraded the building to be a more productive workspace for employees, but it also brought financial benefits to the building owner and tenant. After investing $3.5 million, the owner ended up with a more valuable building. The tenant was thrilled with the ZNE retrofits, because the energy bill fell from $10,000 per month to $0. When developing a new green lease, the building owner asked to raise the rent by $3,000 per month to cover the investment in the building, which the tenant was happy to accommodate, since they were still saving $7,000 per month. Mynt Systems CEO and cofounder Derek Hansen characterized the meeting between the delighted building owner and the pleased tenant as one in which "both of them left the meeting high-fiving each other and clicking their heels." This ZNE project was a win for the climate and a financial win for the building owner and tenant.

Hansen explained that, increasingly, the proposals he submits for commercial ZNE retrofits have irresistible finances. "The last three proposals I did penciled-out to a 3.5-year payback with a 22% return for the real estate investors. These projects aren't just the right thing to do for the environment anymore. They're also good for the bottom line."

From a job creator's point of view, Hansen also notes that his company's line of work creates attractive working-class jobs. "We find that when we hire for construction project manager positions, we don't necessarily need experienced people with advanced degrees to coordinate teams on a construction site. Restaurant shift leaders who have managed several wait staff make great project coordinators. We want to create green jobs for people who don't necessarily have college degrees but yet possess the skills, work ethic, and accountability to learn and grow with the industry."

While companies that do ZNE retrofits can pull talent from existing construction trades, sometimes it's advantageous to recruit people with strong skill sets from other industries.

Jobs needed:

- Construction project managers
- Engineers
- Energy auditors
- Electricians
- Solar installers

- Glaziers
- Roofers
- Finance specialists

PIVOT #6—PART-TIME EMBEDDED SUSTAINABILITY PROJECT MANAGERS

Several years ago, I worked with a candy bar manufacturer whose garbage dumpsters were half full of waste candy bars. Together, we toured the facility looking for opportunities to reduce waste and save money. We started at the beginning of the production process where the ingredients were mixed in a large vat. Then we walked along the extrusion line where a conveyor belt carried a continuous candy bar toward a cutting blade. Next to the machine that cut the extruded candy into individual bars were two big boxes full of waste candy bars. The General Manager explained that the old cutting machine was not functioning well. If the blade landed on a peanut the wrong way, the cut would be lopsided and both sides of the bar would be thrown out.

When I asked the General Manager what ideas he had to reduce candy bar waste, he said they would like to be able to buy a new ultrasonic cutting blade, but it cost $300,000. They could not justify the expense by merely reducing the $28,000-per-year garbage-hauling bill by half; saving $14,000 per year did not provide a fast-enough payback to cover the $300,000 investment.

I then asked if he knew how much money the company spent on ingredients that went into the rejected candy bars. After doing a little research, he calculated the sum to be $300,000 per year, which would provide the necessary payback on the new ultrasonic blade.

Projects like this offer a window into the cost saving opportunities small and medium-sized businesses (SMB) could realize by having a part-time Sustainability Project Manager to work on projects like this. According to the U.S. Census, in 2016 there were 367,446 establishments (facilities) with 100–499 employees: 21,114 in manufacturing, 43,628 in accommodation and food service, and 69,896 in health care and social assistance. In these industries, each facility of this size may be wasting hundreds of thousands of dollars per year on their energy, garbage, water, and supply bills. To save operational costs, these establishments could contract with a consulting firm for a part-time, on-site sustainability project manager a few days per week for six or twelve months to implement projects that would improve the building's operational efficiency.

To develop this type of work, consulting firms would perform an initial evaluation of the SMB's annual operating costs, develop a list of potential operational efficiency projects, and calculate simple paybacks. Projects could include an LED lighting retrofit, HVAC energy efficiency work, solar photovoltaics, landscape water conservation, kitchen and bathroom water conservation, expanded recycling, composting, waste prevention, supply use efficiency, and alternative transportation. From the list of project options, the SMB's CEO or CFO would then choose which sustainability projects they want to pursue, and the consulting firm would embed a sustainability project manager for the amount of time needed to complete those projects. The sustainability project manager would coordinate all the work for these projects, such as supervising independent lighting retrofit contractors, submitting rebate paperwork to utilities, purchasing additional recycling bins and setting up signage, helping staff determine alternative transportation options, and applying for transit subsidies.

From the perspective of the Part-Time Sustainability Project Manager—depending on how many projects the CEO or CFO chooses to implement—they may be able to work at two or three SMBs simultaneously. Then, when the project manager completes the assigned projects, the SMB might choose to hire the project manager to continue working on other projects that add value to the company, or the project manager could move on to the next set of projects at other SMBs.

Sustainability Projects Add Value

Sustainability projects provide three main "buckets" of value: operational cost savings, employee engagement, and green branding. Operational cost savings involve bottom-line financial savings from reducing energy, water, garbage, and supply costs. Employee engagement includes the benefits of being able to attract and retain employees, as well as to optimize their productivity because employees feel their organization cares about their well-being and the well-being of the environment. Green branding yields increased sales from consumers who want to support companies engaged in sustainable business practices.

Which of these three benefits management values most will depend on an organization's culture. Some companies focus on reducing operational costs; other firms put a premium on attracting top talent. Still others feel that having a strong sustainability program differentiates them from others in their industry. I interviewed a colleague at one surgical robotics manufacturing company that uses a rigorous process to find and vet new technical talent, and once they bring people on, management wants their new employees to be happy and productive. The Millennials who predominantly work at this company expect free transit passes, bike racks, recycling programs, excellent lighting quality, and healthy indoor air, and management are happy to oblige since these are what make them more productive employees.

Other companies are more concerned about green branding than employee engagement. One brewery I talked with has no problem finding employees to work there. (Their employees love to be able to say they work at a brewery.) Management's foremost concern revolves around their main competitors, many of whom have strong corporate sustainability programs. They are thus looking at what they should be doing to bolster their own sustainability programs.

Embedded Part-Time Sustainability Project Managers can help SMBs unlock the value that corporate sustainability programs offer without committing to hiring a full-time employee. Doing so also benefits the embedded sustainability project manager who has the opportunity to perform meaningful work for a short amount of time and then move on to the next set of compelling projects.

Jobs needed:

- Sustainability project managers
- Sales and marketing staff

While going through the Living Future Institute's Net Zero Energy Building Certification, Matt Grocoff realized how useful simple, overarching rules are to guide behavior. The certification sets a clear rule that 100% of the building's annual energy needs must be supplied by on-site renewable energy with no combustion allowed. This guides design teams toward a sustainable outcome without over-prescribing how to get there, whether they are in a hot or cold climate.

Although world delegations at the COP21 climate change talks in Paris back in 2015 agreed to limit the global temperature increase to 2°C, and some local governments have Climate Action Plans with greenhouse gas reduction goals, these goals most likely have not yet been incorporated into local building codes. Municipal building departments continue to permit space- and water-heating systems for buildings that burn fossil fuels. Spreading the gospel of a few basic principles like "harvest only what you need and use only what you can harvest" will be invaluable in guiding and empowering citizens to make everyday decisions for a more sustainable future.

CHAPTER HIGHLIGHTS

- Households, businesses, governments, and institutions are spending more on their operational bills than they should. Existing technology is available to make these buildings much more energy efficient and use on-site renewable energy so that buildings can generate as much energy as they use.
- Funding job creation to retrofit single-family homes, multi-family dwellings, commercial buildings, hospitals, and municipal emergency response centers for zero net energy would reduce greenhouse gas emissions.
- Innovative project managers have demonstrated attractive financial paybacks available to those investing in retrofits.

5

MEANINGFUL JOBS FOR

LOW-CARBON MOBILITY

*"Humans are still pretending that fossil fuels
have no probability of a bad outcome."*
—Elon Musk, cofounder and CEO of Tesla

In 2005, in Park Slope, Brooklyn, Marc Wisotsky and his partner Jackie Lew bought two spaces in a parking garage near their home for around $45,000 each. They used one and rented out the other for $600 a month. After taxes and the garage fee, they kept $310. Renting out the space provided a reliable income stream, but the big payoff came when they sold their extra space in 2016 for $285,000.

"We could have gotten more—the prices just keep going up and up," Wisotsky said. "There are never as many parking spaces as residential units being built."[47]

While paying as much for a parking space as a house would cost in many other parts of the country might surprise people, the anecdote provides insight into how much Americans value personal vehicles. New York City boasts abundant public transit options, and yet someone was willing to pay $285,000 for *one* parking space. People treasure personal vehicles partly because many alternative transportation options do not offer frequent or rapid service. If we as a society invest in low-carbon mobility alternatives to personal vehicles, we will create viable alternatives and reduce many of the problems caused by internal combustion engines.

We know that burning fossil fuels contribute to poor air quality and climate change.

FIGURE 16: TOTAL U.S. GREENHOUSE GAS EMISSIONS BY ECONOMIC SECTOR IN 2016

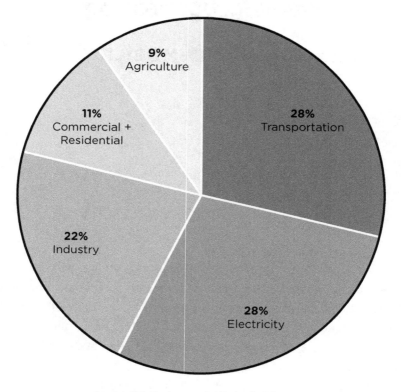

Source: U.S. Environmental Protection Agency

Let's set aside the knowledge that the U.S. transportation sector combusts 391 million gallons of gasoline and diesel fuel every day (and that doing so contributes to respiratory illnesses such as asthma), as well as the fact that the transportation sector generates 28% of U.S. greenhouse gases (which contribute to climate change) as seen in Figure 16.[48] Instead, let's focus on the other major compelling reasons to transition to alternative transportation options. As we think about what we want our future transportation system to look like and do, and what jobs are needed to build this system, having a solid grasp on the problems our current transportation system causes will help us make the case for creating these new transportation jobs.

PARKING TAKES UP VALUABLE REAL ESTATE

In a 2012 article, *New York Times* journalist Michael Kimmelman tried to pin down the number of parking spaces in the U.S. After reviewing various studies, the closest number he could arrive at was a range of 105 million to two billion parking spaces. One study claimed the U.S. has eight parking spaces for each vehicle; another finding stated that Houston has 30 parking spaces for every resident.[49]

Regardless of the exact number, dedicating so much real estate to parked vehicles comes at a high opportunity cost. Assuming an average parking space takes up 330 square feet,[50] this impervious pavement will not be available for important concerns like nature, pedestrian walkways, and housing.

PERSONAL VEHICLES ARE EXPENSIVE

While parking in most parts of the country is free, the vehicles themselves are expensive and can consume a large share of household income. Most of us don't realize the sum total we're spending on our vehicles every year, with payments for various vehicle expenses going to different companies and agencies. Personal finance website NerdWallet calculated that in 2017, a personal vehicle driven 15,000 miles per year costs the owner an average of $8,469 per year. This includes an average car payment of $523 per month—as calculated by credit reporting agency Experian—and gas, insurance, maintenance, repairs, registration, fees, and taxes.[51] People who own their cars and don't have a car payment spend less per year but still pay thousands of dollars simply to keep it in their possession.

When looking at the costs of personal vehicles, some studies also add in the personal cost of lost productivity when sitting in traffic. Assuming an average wage of $26 per hour, five hours per week of lost productivity costs the average commuter $6,760 of their time each year.[52]

SINGLE OCCUPANCY VEHICLES CAUSE TRAFFIC CONGESTION

Speaking of lost productivity, housing developments located far from job centers create some horrendous commutes in many metropolitan areas. The Auto Insurance Center created a list of the top-ten cities with the highest percentage of people commuting over one hour to work, as shown in Table 2.[53]

The suburbs and exurbs of Washington, D.C., New York City, San Francisco, and Los Angeles suffer the longest commute times, but many other urban areas around the country also struggle with traffic congestion.

TABLE 2: TEN WORST COMMUTES IN THE U.S.

NO.	CITY	DETAILS
10	Chino Hills, CA	Average commute time: 37.3 mins Percentage of people who spend 60+ minutes commuting: 20.7% Population: 76,796
9	Hesperia, CA	Average commute time: 37.3 mins Percentage of people who spend 60+ minutes commuting: 22.3% Population: 92,309
8	Hoboken, NJ	Average commute time: 38.9 mins Percentage of people who spend 60+ minutes commuting: 14% Population: 52,452
7	New York City, NY	Average commute time: 39.9 mins Percentage of people who spend 60+ minutes commuting: 25% Population: 8.4 million
6	Pittsburg, CA	Average commute time: 40.2 mins Percentage of people who spend 60+ minutes commuting: 29% Population: 66,947
5	Dale City, VA	Average commute time: 40.4 mins Percentage of people who spend 60+ minutes commuting: 26% Population: 72,130

NO.	CITY	DETAILS
4	Palmdale, CA	Average commute time: 40.7 mins Percentage of people who spend 60+ minutes commuting: 31.8% Population: 156,672
3	Tracy, CA	Average commute time: 40.8 mins Percentage of people who spend 60+ minutes commuting: 31.7% Population: 85,284
2	Antioch, CA	Average commute time: 42.5 mins Percentage of people who spend 60+ minutes commuting: 31.1% Population: 107,501
1	Waldorf, MD	Average commute time: 43.4 mins Percentage of people who spend 60+ minutes commuting: 31.8% Population: 71,399

Source: Auto Insurance Center

PERSONAL VEHICLES ARE DANGEROUS

Personal vehicles are convenient, but injuries and deaths from traffic accidents are part of the price we pay for that convenience. In 2017, 40,100 people lost their lives in motor vehicle accidents in the U.S.[54] Factors that contributed to vehicle deaths include distracted driving, drunk or drug-impaired driving, lack of seat belts, speeding, and drowsy driving.

When stepping back and taking a clear-eyed look at the problems posed by our current transportation system, it's easy to think we could do better. Given the amount of land lost to parking spaces, the expense of personal vehicles, the lost productivity suffered in traffic jams, the injuries and lives lost in accidents, the respiratory health effects, and transportation's contribution to climate change, building out a mix of low-carbon transportation alternatives would yield many benefits for people and the environment. In the course of building a new, healthier transportation system, we will create meaningful work for many people.

Investments are already flowing into alternative transportation options, including EVs, EV charging infrastructure, autonomous vehicles, dockless e-bikes and e-scooters, and commute management platforms, all of which would result in major changes to the transportation sector in the next decade. The big unknown is how much of a role autonomous vehicles will play in the future of mobility.

DISRUPTION IS COMING

In April 2018, the former Vice Chairman of General Motors, Bob Lutz, spoke at the annual meeting of the Society of Automotive Engineers in Detroit with unwelcome news. "We are approaching the end of the automotive era," he said. He predicted that vehicle autonomy will gradually take over more and more of the task of driving, and that "the end-state will be the fully autonomous module with no capacity for the driver to exercise command. You will call for it, it will arrive at your location, you'll get in, input your destination and go to the freeway." He continued that "people who like to drive, and car companies that rely on branding, have another 25 years at the most. After that, it's all over."

Lutz admitted that this transition will bring "a vast improvement to national productivity." Traffic jams and accidents will become rare events, and perhaps some 90% of the 40,000 annual traffic deaths in the U.S. will be eliminated.[55]

The Big Three U.S. auto manufacturers anticipate an imminent shift in mobility. They know that the average personal vehicle sits unused 95% of the time,[56] that Millennials buy vehicles less often than previous generations, and that internal combustion engines contribute to climate change. They understand these problems and are planning for a different future. As well, because the Big Three know that people want mobility but not necessarily to own a vehicle, they anticipate a post-ownership future.

This is why automakers have been investing in Mobility-as-a-Service and self-driving vehicles like the ones Lutz describes: because they want to be part of the future of smart, low-carbon mobility. Ford invested in bike-sharing systems; GM runs a car-sharing service called Maven; and Waymo purchased 62,000 Chrysler minivans to test self-driving technologies. All three are hedging their bets in order to figure out how the smart mobility future will shake out.

HOW BIG OIL WILL DIE

As auto manufacturers prepare for a shift to electric autonomous vehicles to maintain market share, people who care about climate change are pinning their hopes in part on autonomous vehicles as a means to shift away from fossil fuels. Many of us, frustrated by the slow pace of divestment from fossil fuels by institutional investors, gleefully read Seth Miller's 2017 *NewCo* article "How Big Oil Will Die."[57] The crux of Miller's convincing argument revolves around the potentially dramatic cost savings as society switches from internal combustion engines to electric vehicles, then automates the driving function.

Consider the following technical comparison of internal combustion engines and electric vehicles, according to Miller:

- An internal combustion engine drivetrain has about 2,000 parts
- An electric vehicle drivetrain has about 20 parts
- A system with fewer moving parts will be more reliable
- Internal combustion engines last about 150,000 miles,[58] whereas current estimates of the lifetime of electric vehicles is 500,000 miles[59]

Besides improved reliability and a longer lifespan, the total cost of owning an electric vehicle is a fraction of the cost (one-quarter to one-third) of a gasoline-powered vehicle. Miller cites Stanford University professor Tony Seba's analysis comparing the 2017 rate of $0.535 per mile that the Internal Revenue Service (IRS) allowed businesses to write off their taxes, versus the $0.13 per-mile cost for a fleet electric vehicle. For those in New York City who drive a taxi 50,000 miles per year, a $0.40-per-mile saving adds up quickly. Over the course of a year, a taxi driver could save $20,000.

Now consider that a taxi ride in New York City costs about $2.00 per mile. That ride, which would be priced at $1.60 per mile for an electric vehicle, costs significantly more than the $0.535 per mile the IRS lets business taxpayers write off for a vehicle they own. This is because the most expensive part of the cab ride is the driver. By removing the driver from the equation, the cost per mile of an autonomous electric vehicle is dramatically cheaper than an internal combustion engine vehicle with a driver.

Seba predicts the following roll-out of autonomous vehicles:

- Self-driving cars will launch around 2021
- A private ride will cost $0.16 per mile, eventually falling to $0.10
- A shared ride will cost $0.05 per mile, eventually falling to $0.03
- Oil use will peak by 2022
- Used car prices will crash by 2023 as people give up their vehicles; new car sales for individuals will drop to nearly zero
- Gasoline consumption by cars will drop to near zero by 2030, and total crude oil use will have dropped by 30% over current numbers

The aggressive timeframe in this scenario seems like wishful thinking for those who would like to see autonomous vehicles succeed and the oil industry descend into a death spiral. Even if this transition rolls out more slowly, the driving force behind it is the main assumption in Seba's analysis: given a choice, people will select the cheaper option.[60]

When I asked Miller about whether the adoption of autonomous vehicles will work in rural areas, as well as urban and suburban areas, he responded that the process may look a lot like the growth of cell phone coverage: cell phone companies first expanded coverage in urban and suburban areas, then gradually extended coverage beyond, but some rural areas stayed off the map for quite a while. He explained:

"I think AVs will work in rural areas eventually, just like Verizon does (finally) now. In rural areas, it will be harder to sell your existing vehicle and count on transportation-as-a-service full time, but existing vehicles will be able to stay in the garage more often. The sick, the elderly, and the drunk will all create plenty of demand on their own, and it will grow from there. The first rural use cases may not be the ones where AV is cheaper; they will be where driving yourself is not an option at all."

People adopt new technologies for many different reasons. As we think about what a low-carbon mobility future looks like and the work needed to build it, let's consider the human benefits that will accrue: cost savings, improved health, improved quality of life, and a stable climate.

At the same time, we need to plan for the disruptions that will happen to the working world as we transition from one dominant transportation method to several different options. Because in the next decade potentially millions of people who drive for a living could suffer the hardship of job loss, we need to draft a strategy for their transition into new lines of work.

A GIANT SHIFT IN TRANSPORTATION JOBS

In 2016, 4,285,400 people in the U.S. derived their primary income from driving. According to the Bureau of Labor Statistics, the categories for professional drivers break down into the following numbers:

- Heavy truck and tractor trailer drivers—1,871,700[61]
- Delivery truck drivers—1,421,400[62]
- Bus drivers—687,200[63]
- Material moving machine operators—682,000[64]
- Taxi, ride-hailing drivers, and chauffeurs—305,100[65]

In addition to the people who drive for a living, U.S. car and truck dealerships employed 1.1 million people in 2016, according to the National Automobile Dealers Association.[66] This includes those who sell and service vehicles. Will a large number of these drivers and people who sell and service vehicles need to transition to other work? If so, what work will be available for them?

The jobs available in the transportation sector of the future will depend partly on two things: our vision for our future transportation system and how we invest in it over the next decade. Do we value having well-developed alternative transportation options that are safer, healthier, cleaner, cheaper, and better for the climate? This is a discussion worth having in our society. Engaging in civic discussions about the future of our transportation system and encouraging specific types of public and private investments will ensure we help steer the discussion, not get run over by it.

PIVOT #7—DEVELOPMENT OF MASS TRANSIT

When I ride on high-speed trains in Japan, China, France, Spain, Italy, and Germany, I feel pangs of jealousy. These countries invested in high-speed rail and as a result have fast, quiet, intercity service. As an alternative to flying or driving between major cities, people in these countries can take trains that travel at speeds of around 200 miles per hour.

The U.S. Department of Transportation has envisioned a network of high-speed rail lines that could connect people between metropolitan areas.

FIGURE 17: POTENTIAL ROUTES FOR HIGH SPEED RAIL

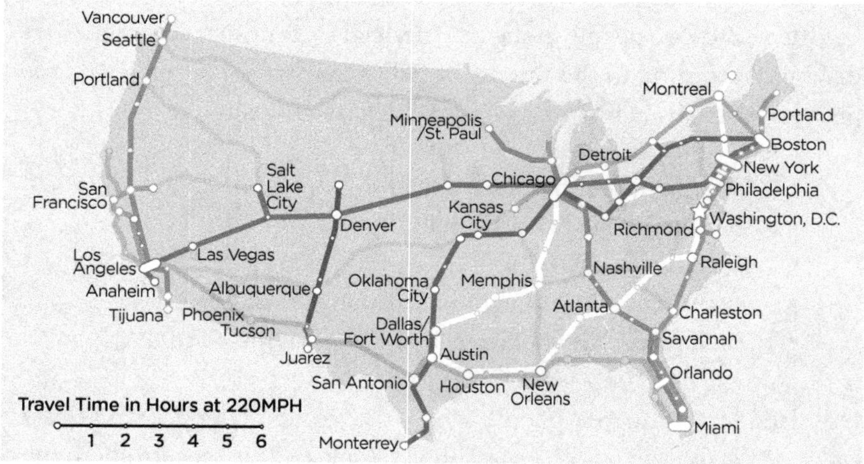

Source: U.S. Department of Transportation

Investing in high-speed rail at the same level we have invested in the interstate highway system and airport infrastructure would yield a viable interregional rail system. According to the map above, a high-speed train travelling between downtown Denver and downtown Los Angeles would take about six hours. A non-stop flight on Southwest Airlines is listed as two hours and 25 minutes, but between driving to the airport, parking, checking baggage, waiting in security lines, and waiting to board, then travelling to one's ultimate destination, the trip can easily take six hours, door to door.

The U.S.'s mass transit options at the interregional, intercity, and local levels are anemic. Retrofitting our auto-friendly transportation infrastructure is a challenge, but it is possible to overlay additional mass transit options to move more people faster than single-occupancy vehicles. Laying new commuter train lines and light rail can work for suburban areas, but for more densely developed areas, bus rapid transit (BRT) offers the most cost-effective retrofit option.[67]

BRT's five basic features show how it is different from a regular city bus:

- **Dedicated lanes**—Right-of-way lanes ensure buses are never delayed, because they avoid mixed-traffic congestion
- **Bus alignment**—A center-of-roadway travel corridor keeps buses away from busy curbside areas where cars park, stand, and turn
- **Off-board fare collection**—Fare payment at the station instead of on the bus speeds up boarding
- **Intersection modifications**—Prohibiting turns for traffic across the bus lane reduces delays
- **Platform boarding**—Riders have already paid before they board and can walk straight onto the bus from the platform, eliminating stairclimbing (which slows boarding for everyone) and making boarding easier for people in wheelchairs, the disabled, and people with strollers[68]

BRT takes a full lane away from single-occupancy vehicle traffic each way on major thoroughfares but is worth the sacrifice. With thoughtful transportation planning, BRT will move many more people along major streets than personal vehicles.

Jobs needed:

- Transportation planners
- Construction project managers
- Construction workers
- Finance specialists

PIVOT #8—TRANSPORTATION MANAGEMENT ASSOCIATIONS

When I started working as the sustainability manager at NASA Ames Research Center, I drove to work alone in 40 minutes. I knew I should be taking alternative transportation to work but was unsure of the best mix of commuting options. I experimented. The first day I tried bicycling and taking the train, the trip took me two hours each way. The fact that this commute took so long was heartbreaking. I had a small child at home and could not afford to spend an extra two hours and 40 minutes commuting each day.

Over a few successive days, I tried different combinations of trains, buses, shuttles, bicycling, and walking and reduced the commute to one hour and twenty minutes each way. Although this took longer than driving, I committed to this commute with a fourteen-minute train ride and an eight-mile bike ride, each way, twice a week. The sixteen-mile, roundtrip bike ride was my workout for the day, meaning I didn't need to go to the gym that day, and we had showers at work so I could freshen up. The avoided extra gym and commuting time were equivalent so, ultimately, the alternative commute did not come at the expense of my family.

While figuring out the best mix of options requires some trial and error, other commuters may not be as dedicated and persistent. This is why we need more Transportation Management Associations (TMA): to help those willing to undertake alternative commutes find the optimal solution. Abundant free parking and passing the time on long commutes with music, news, or podcasts can seem irresistible, so in the face of the ease and convenience of driving, enticing people to shift to different transportation modes feels like a lot of heavy lifting.

TMAs ease that burden by working to understand and then eliminate barriers to shifting transportation modes. Their staff serve as transit concierges, helping people figure out the best alternative transportation options available to them. Once the TMA has presented key information about the cost, duration, and financial incentives of alternatives to driving, people then do an internal calculus to determine if the perceived benefits of the mode-shift outweigh the perceived barriers of the change in lifestyle. Part of Transportation Demand Management work involves spreading the message to commuters within a given organization or area that alternatives exist. Sometimes, just asking helps. In one TMA, where I conducted outreach in a target business district about an upcoming mode-shift project, one person said, "You know, I should bike to work. I only live three miles from here. I know I should, and I will start doing that."

Stanford Research Park (SRP) near Stanford University in Silicon Valley has five full-time TMA staff members. Staff use several different methods to encourage the 29,000 SRP employees to consider taking a train, bus, shuttle, vanpool, carpool, bicycle, or electric scooter or to walk or even telecommute. SRP's TMA conducts events, such as monthly bike-to-work days, cycling safety classes, e-scooter testing, and ice cream socials. Other activities the TMA staff lead to educate, inspire, and incentivize people to consider alternative transportation include:

- Writing and distributing a regular newsletter
- Carpool matching
- Running a shuttle to nearby train stations
- South Bay commuter bus
- Running vanpools
- Coordinating a $100/month incentive program
- Cultivating bike and transit champions
- Building community in person and on social media

Each year, the SRP conducts a commuter survey to understand where employees are coming from, how they get to work, and why they do or do not use alternative commute options. Their 2017 commuter survey of 6,000 workers found that 68% drive alone and 32% either take alternative commuting options or telecommute, compared to 73% and 27% in 2016 and a nation-wide 77% drive alone rate.[69]

Sometimes, it's the little things that keep people from following through on their ambition to shift modes. In mild climates like the Bay Area, many more people would like to bicycle to work but worry about not looking presentable in the office. Electric-assist bicycles are a good option for people who live five to ten miles away: they can travel to work without working up a sweat and then pedal home to get exercise. To those who avoid bicycling for fear of "helmet hair," abundant advice proliferates online about how use beach spray, scarves, braids, and dry shampoo to

combat this problem. These are just some of the little, perceived imped-
iments that trip people up and that TMAs can help people overcome.

Financial incentives help, too. The Palo Alto TMA, near the Stanford Research Park, offers free transit passes to people who work in
downtown restaurants, shops, and hotels. Eighty percent of the people
who work in the downtown service sector used to drive, because they
already had a car and could not possibly afford a transit pass on top of
the expense of a car. With revenues from downtown parking garages and
residential parking passes, as well as donations from large employers in
the area, the Palo Alto TMA gave away free transit passes and reduced
the rate of single-occupancy vehicle driving by downtown service sector
employees from 80% in 2016 to 56% in 2018.

Many low-income service sector workers who receive these passes
have expressed deep appreciation for the program, such as:

- "I'm thinking of going back to school with the money I
  save by not driving."—Maria
- "I am the TMA's biggest fan. I cherish this transit pass."—Tim
- "I used to bike all the time, but I haven't been lately. Now
  I bike to the train station and take the train to work. I
  forgot how much I like biking. Next weekend, I'm going
  to bike up to San Francisco."—Cristian
- "The train is so much faster than driving, and I get to
  relax on the way to work."—Romina

Once commuters tried alternative transportation, they realized
the benefits.

Another major way TMAs add value is by helping people figure out
the first mile/last mile issue. Many people could get to work by taking
one train or bus but live or work more than a walkable distance from the
transit stop; they are just not sure how to close the gap. Thus, TMAs
are experimenting with transportation rental options such as bicycle-,
scooter-, and car-sharing systems in various cities. Having a mix of dockless vehicles that users unlock via mobile app could be the solution to
the first mile/last mile problem, but further study and experimentation
are required prior to full-on implementation.

From a TMA's perspective, behavior change management is not a "one and done" effort. Mode-shift outreach must happen on a regular, recurring basis as employees in organizations leave and new ones start. For towns, large business parks, and organizations with a campus in a major metropolitan area that struggle with traffic congestion and parking problems, having a TMA is a way to help people explore alternative transportation options and figure out which mix of options work best for them.

Jobs needed:

- Project managers
- Outreach coordinators

PIVOT #9—DEVELOPMENT OF SAFE BICYCLING INFRASTRUCTURE

Upgrading roadways to create protected bike lanes would go a long way toward making people feel safer bicycling on roads: a survey of 16,000 Americans by People for Bikes found that 53% would like to bike more but are concerned about being hit by a car. It also found that 46% would be more likely to ride more often if bicycles were physically separated from cars.[70]

The beauty of protected bike lanes is that distracted drivers cannot accidentally run into cyclists. Part of the reason more cities don't adopt protected bike lanes is the widely circulated statistic that they cost $1 million per mile. In response to this misinformation, People for Bikes published a blog post, "No, Protected Bike Lanes Do Not Need to Cost $1 Million Per Mile,"[71] explaining several different protected bike lane options and their cost ranges:

- Parked cars—$8–16k/mile
- Oblong low bumps—$10–20k/mile
- Delineator posts—$15–30k/mile
- Turtle bumps—$15–30k/mile
- Large bumps—$15–30k/mile
- Parking stops—$20–40k/mile
- Linear barriers—$25–75k/mile

- Cast-in-place curb—$25–80k/mile
- Jersey barriers—$80–160k/mile
- Planters—$80–400k/mile
- Rigid bollards—$100–200k/mile
- 12" precast curb—$400–600k/mile

Retrofitting cities for protected bike lanes does require tasks other than bike lane installation by contractors or public works staff. Another key ingredient is community outreach: since there may be community pushback about the loss of on-street parking and changes to traffic patterns, local city planning staff, along with relevant organizations, must explain proposed streetscape changes to the public, receive community feedback, and make design changes accordingly. This may become less of an issue where and when autonomous vehicle availability obviates the need for personal vehicles (which require street parking) and opens up more space on streets for protected bike lanes.

As an alternative to installing protected bike lanes, some cities have chosen to make traffic engineering changes to reduce the speed of vehicles and thus make the roads safer for bicyclists. Data show that when a vehicle–bicycle or vehicle–pedestrian crash occurs, a lower vehicle speed greatly increases the cyclist or pedestrian's chance of survival.[72] Figure 18 shows how effective traffic calming measures can be.

FIGURE 18: PEDESTRIAN'S CHANCE OF SURVIVING A COLLISION WITH A VEHICLE

A Pedestrian Hit by a Vehicle Traveling at
25 MPH
Survivability
Has an **89%** Chance of Survival

A Pedestrian Hit by a Vehicle Traveling at
35 MPH
Survivability
Has a **68%** Chance of Survival

A Pedestrian Hit by a Vehicle Traveling at
45 MPH
Survivability
Has a **35%** Chance of Survival

Source: B. C. Tefft, "Accident Analysis & Prevention"

Reducing vehicle speeds by as little as ten miles per hour makes a significant difference to vulnerable road users like bicyclists and pedestrians.

One traffic re-engineering measure that makes roadways safer for cyclists has to do with traffic intersections. Figure 19 shows how many conflict points exist in a typical intersection and in a roundabout.

FIGURE 19: NEIGHBORHOOD TRAFFIC SAFETY

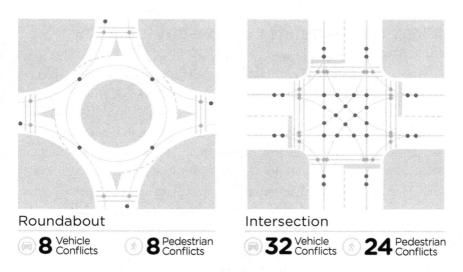

Roundabout

8 Vehicle Conflicts   **8** Pedestrian Conflicts

Intersection

32 Vehicle Conflicts   **24** Pedestrian Conflicts

Source: Transportation Division, City of Palo Alto, California

Roundabouts force traffic, whether cars or cyclists, to head in the same direction. Conflicts between vehicles and bicycles with traffic flowing in the same direction have better outcomes than vehicles and bicycles heading toward each other or crossing at 90° angles.

To make city streets safer for cyclists, municipal public works departments need enough staff and contractors to plan re-engineering changes, work with the community on those proposed changes, and retrofit roadways. Doing so will reassure the 99% of American workers who do not regularly bike to work[73] that they will be safe on the roadways.

Jobs needed:

- Transportation planners
- Construction project managers
- Construction workers
- Community outreach specialists
- Bike manufacturing
- Bike repair specialists

PIVOT #10—MOBILITY-AS-A-SERVICE APPS

Imagine that in a few years, you won't own a car. Instead, you will rely on the numerous iPhone and Android apps that help people plan and pay for a multi-modal trip. You might use Moovel, Moovit, Urban Engines, Transit App, TripGo, TransLoc, Swiftly, Citymapper, VentraUS, Siemens, Whim, or TravelSpirit. Let's say that in a few years you prefer to use Moovit. You are in San Francisco and have a client meeting in Mountain View. Without any (active) input from you, Moovit checks your Google Calendar, plans your trip, and picks the best option based on your trip length, budget, and comfort preferences: LyftLine to Caltrain to a private microtransit van.

Moovit then pre-books a scarce seat on the twelve-seat van, schedules the Lyft trip so that you make the train on time, then prods you to pack up your bags so you're ready for pickup. You pay for Caltrain by tapping your phone, then take a nap until Moovit wakes you just before reaching the station. On foot, you get lost between the train station and the microtransit pickup spot, but Moovit's there to guide you toward it. Your employer pays for half of the trip, and while there's Lyft/Caltrain transfer discount involved, both price adjustments occur seamlessly in the background. You arrive at your client site oblivious to the artificially intelligent Moovit agent that made every single aspect of this trip effortless for you, all using only one app.

Transportation planning and policy expert Steve Raney conceived of this scenario to describe a vision more Mobility-as-a-Service (MaaS) apps could be working toward. Various apps provide separate, functional pieces of this vision, but none pull them all together. Ventra by

Cubic, for example, serves Chicago-area commuters who want to plan, manage, and pay for their journeys across three transit systems with their mobile devices. Swiftly provides real-time arrival and departure data for public transportation to improve urban mobility. Commuter management system RideAmigos allows employees to plan and track alternative transit trips and to earn incentives for avoiding single-occupancy vehicle trips. Luum integrates parking management, transit pass administration, and employee communications on a commute data platform. Lyft and Uber help solve the first- and last- mile issues for transit riders with car rides, bikeshare, or scooter share options. People may want to take a commuter train but live two miles from the station—and at the other end of their journey, their ultimate destination is a full mile from the train station. Each of the companies mentioned provide a part of the solution, which remains fragmented. MaaS apps can put it all together.

MaaS apps will continue to evolve as the industry offers various features, acquires feedback, and pivots to meet the public's needs. At some point, MaaS will ideally be able to help people navigate the optimal combination of options among trains, buses, ridesharing, vanpools, shuttles, electric bicycles, bikeshares, personal bicycles, electric scooters, and walking. As more investments are made into their development, MaaS could help a larger percentage of the public shift away from using internal combustion engines and help society lower the transportation sector's footprint.

Jobs needed:

- Project managers
- Software engineers
- Marketing specialists

PIVOT #11—ELECTRIC VEHICLE CHARGING INFRASTRUCTURE

Every year, EV offerings improve. Longer driving ranges, more powerful acceleration, and great design features entice more and more people to test drive EVs. Federal and state government tax incentives and rebates to purchase electric vehicles then seal the deal.

What's holding back large-scale adoption at this point is the paucity of electric vehicle charging infrastructure. Everyone has Level 1 (120 volt) chargers at home; these are regular outlets. We need more Level 2 (240 volts) electric vehicle chargers as well as Direct Current Fast Charging (DCFC) stations. Prioritizing the build-out of EV charging infrastructure in the following five areas will reassure the public that chargers will be available when they need them at:

- Multi-family complexes (Level 2)
- Workplaces (Level 2 and DCFC)
- Transit parking facilities (Level 2)
- Popular destinations such as shopping centers, arenas, parks, recreation areas, etc. (DCFC)
- Near transportation corridors such as highways (DCFC)

First and foremost, people need to be able to charge their EVs at home and at work. People with EVs recognize that most of their trips happen between home and work, and most people do not commute more than 40 miles. EVs therefore provide a perfect fit for commuting and running errands around town.

Lately, auto manufacturers have released EVs with longer and longer ranges, allowing intrepid EV drivers to venture farther and farther from home. Several EVs have ranges of over 200 miles, allowing people to take their EVs on vacation. Longer-range EVs encourage development of charging stations in public spaces and along transportation corridors.

Municipalities are key in encouraging people to choose an electric vehicle when shopping for their next car. Several steps municipalities can take to reassure people with "range anxiety" around EVs are:

1. Strengthen building codes to require EV charging infrastructure installation for new buildings and renovations, with a density of one charger per residential unit
2. Allow private companies to finance and operate charging stations in areas with public access, for example shopping malls and gas stations, to encourage the availability of DCFC infrastructure for long-distance travel

3. For residential and workplace charging, encourage building owners to apply for grants and utility-funded installations
4. Encourage public signage for EV charging stations that is visible from the nearest public roadway
5. Conduct "EV Ride and Drives" and related educational activities to familiarize people with the advantages of EVs
6. Pilot municipal codes requiring some level of EV charging infrastructure for existing multi-unit dwellings and workplaces[74]

These measures will cost municipalities little or nothing to implement beyond staff time but will help accelerate deployment of EV charging infrastructure.

Those working on expanding charging infrastructure must also determine how to find funding, but options exist: grants from utilities and government agencies may be available in some regions to help reduce the cost, and public-private partnerships also help make projects financially viable. The City of Palo Alto, for example, received grant funding from the Bay Area Air Quality Management District to pay for part of the cost of their recent EV infrastructure project, then contracted with Komuna Energy to build, own, and operate 1.3MW of solar photovoltaics and install dozens of EV charging stations at four city-owned parking structures. ChargePoint manages the charging system in which people who want to charge their EVs at these public parking garages pay $0.23/kWh, about $2 for a charge.[75]

This type of public-private partnership showcases the best of what each sector brings to the table. The public sector contributes public space and grant money, while the private sector invests private capital, does the installation, and runs the charging stations.

There's much more work to be done to build out a vast network of EV charging stations at homes, multi-family complexes, workplaces, and in public spaces. Developing an extensive network is vitally important if we are to lower the carbon footprint of the transportation sector. A positive feedback loop is in play: the more EV charging stations people see, the more likely non-EV owners who are considering buying an EV will be reassured that charging stations will be available when they need them and thus will make the leap.

Jobs needed:

- Municipal transportation planners
- Construction project managers
- Electricians
- Finance specialists

PIVOT #12—DESIGNING WALKABLE COMMUNITIES

A key part of reducing the carbon footprint of the transportation sector is redesigning communities so that people do not need to get in a car every time they want to go somewhere. Many people would prefer a more pedestrian-friendly lifestyle. A 2017 National Community Transportation Preference Survey found that 62% of Millennials and 66% of the Silent Generation would rather be able to walk to shops and restaurants, even if it means living in an apartment or condominium.[76] Contrast that with Baby Boomers and Generation Xers, of whom a larger percentage prefer car-dependent living in detached single-family homes to walkable communities and living in apartments or condos.

For those who would like more walkable communities, WalkScore rates cities with more than 200,000 residents. Users in the 141 cities WalkScore rates can type in the address of a place they are considering moving into and find a score on a scale of 0 to 100 for walkability, transit-friendliness, and bikability. A high-end score describes a walker's paradise, and a low-end score describes a place where most errands require a car. Walk Score measures the distance from specific addresses to amenities in several categories. Amenities within a five-minute walk (0.25 miles) are given maximum points. Fewer points are given to more distant amenities, with no points given for locations over a 30-minute walk. The Transit Score rates the usefulness of nearby transit routes based on frequency, type of route (rail, bus, etc.), and the distance to the nearest stop on the route. For this metric, a score closer to 100 is more transit friendly. Finally, the Bike Score measures bike infrastructure elements such as bike lanes and trails, as well as hilliness, destinations, road connectivity, and typical number of bike commuters (in order to judge the "social" aspect of a route). Locations with a score closer to 100 are more bike friendly.

One example of a town with a high Walk Score, Transit Score, and Bike Score is Hoboken, New Jersey. Right across the Hudson River from New York City, Hoboken boasts a population of about 50,000 people and a vibrant main street filled with shops, restaurants, cafes, and bars. Above many of these street-level amenities sit a few stories of apartments. Frequent trains and buses provide easy access to jobs in New York City and other parts of New Jersey. Because of these factors, Walk Score gives Hoboken, NJ, a 95 Walk Score, a 75 Transit Score, and a 70 Bike Score. Rents are high in Hoboken compared to other parts of the country, but the trade-off is that people don't need a car, which saves them money on car payments, gasoline, maintenance, repairs, parking, and other associated expenses. By showcasing spaces like Hoboken with high walkability, transit-friendliness, and bikability scores, Walk Score hopes to encourage other cities to make their communities more walkable.

For towns and cities that could use a little community engagement to help make their public spaces more pedestrian-friendly, Dallas-based non-profit Better Block Foundation offers an innovative model. The project was launched in April 2010 with the aim to bring together community organizers, neighbors, and property owners to improve blighted downtown areas with vacant properties, wide streets, and few amenities. Initially, IT consultant and cycling advocate Jason Roberts and transportation planner Andrew Howard envisioned the transformation of a four-block center in the historic downtown Oak Cliff neighborhood of Dallas. With little money and no permanent infrastructure but abundant community support, they set out to design and implement the revitalization within 24 hours.

Roberts and Howard invited several friends to build temporary awnings, bike lanes, café tables, and medians with potted plants out of plywood. The idea behind these guerrilla street redesign tactics is to temporarily assemble pieces that help community members envision what the area could look like. To maximize community engagement, Roberts and Howard recruited local artists and food vendors to sell goods inside and outside abandoned storefronts. Then the community decided which pieces worked best and should be permanently installed.

The whole project was a true learning experience for the organizers. They sought to engage the community in figuring out what was valuable to them and what amenities could help them reclaim blighted areas. Going into the project, Howard admitted that they were convinced they would be met with bureaucratic resistance. He also said that with this guerrilla street revitalization, "We broke as many laws as we could. We were ready to go to jail."[77] The project organizers refer to this unregulated urban planning approach as "Urban Defibrillators," with the intent to provide the "shock" cities need.

Better Block provides a methodology for jumpstarting urban redesign of neglected downtown areas and creating spaces people appreciate and use, as well as provide people in the community the tools and resources necessary to make the upgrades. This project empowers ordinary citizens to make simple changes at the block level, which can create a dramatic ripple effect through the city. They also show how making small but impactful changes can greatly reduce our impact on the climate while creating meaningful opportunities and jobs and fostering local business.

Word about the Better Block project spread. In 2017, Ottumwa, Iowa, invited Better Block to convene community members for an eight-hour visioning and design session.[78] From their vision, they created four pop-up shops, a playground on a vacant corner lot, street furnishings, a bike rack, a protected bike lane, colorful crosswalks, curb extensions, landscaping, and public art. The project was topped off with a concert held in a long-vacant downtown theater building attended by 500 community members. Residents loved the temporary revitalization and are considering the next phase.

For other communities that want to try something similar, Better Block offers an open-source toolkit of designs for benches, chairs, planters, stages, bus stops, beer garden fences, and kiosks. People can take these designs, cut them out of plywood, and assemble them in their target areas.[79]

Jobs needed:

- Urban planners
- Community organizers
- Urban infill developers
- Construction workers

CHAPTER HIGHLIGHTS

- Owning a personal vehicle costs $8,000–$9,000 per year between car payments, gasoline, maintenance, repairs, insurance, and license and registration.
- Over four million people currently drive for a living in the U.S. and 1.1 million work in vehicles sales and service.
- Disrupting our high-carbon mobility system with low-carbon mobility projects could create jobs for people displaced by autonomous vehicles in the near future.
- Projects to build out a low-carbon mobility system include developing mass transit, creating transportation management associations, implementing safe bicycling infrastructure, developing Mobility-as-a-Service apps, installing electric vehicle charging infrastructure, and designing walkable communities.

6

MEANINGFUL JOBS FOR

A CIRCULAR ECONOMY

"The green economy should not just be about reclaiming throw-away stuff. It should be about reclaiming thrown-away communities. It should not just be about recycling things to give them a second life. We should also be gathering up people and giving them a second chance."
—Van Jones

On a 2013 diving trip off the coast of Greece, sixteen-year-old Boyan Slat became concerned when he saw more plastic in the water than fish. As his mind jumped from one possible cleanup solution to the next, the Dutch teenager wondered if a curved barrier could use the power of ocean currents to corral and aggregate drifting plastic debris.

That idea came into fruition in 2018 when a solar-powered, 2,000-foot-long, floating pipeline with a ten-foot-deep skirt was launched from San Francisco. Armed with $20 million in funding, the mission of Slat's Ocean Cleanup project is to clear a portion of the estimated 1.8 trillion pieces of plastic comprising the Great Pacific Garbage Patch, an area of the ocean between California and Hawaii twice the size of Texas.[80]

"What we're trying to do has never been done before. So, of course, we were expecting to still need to fix a few things before it becomes fully operational," Slat explained. Four weeks into testing, the U-shaped device was capturing plastic but then losing it. A team of experts is now working on a possible fix,[81] and Slat dismissed critics. "Big problems require big solutions," he said. "If anyone has any better ideas, we'd love to know."

The Great Pacific Garbage Patch is just one zone in the world's oceans where plastic accumulates. Efforts to remove plastic from the oceans should continue, but the true long-term solution is to redesign our disposable, single-use economy and transition to a circular economy. The fact that 99% of the things we buy in the U.S. become waste within six months[82] is evidence of the need for an economy that generates less waste. While the U.S. boasts sophisticated waste management systems, the sheer volume of single-use disposables overwhelms our ability to contain it all. The good news is that building and running a circular economy will create jobs for people, some of which are well-suited for those with entry-level skills.

A circular economy is one that keeps resources in use as long as possible, thereby extracting their maximum value. Then, at the end of each service life, products are recovered and products and materials are regenerated.[83] The transition starts by changing how we think about consumer products: redesigning for durability and reuse, sharing items that sit idle most of the time, and upcycling items that still offer utility. Ideally, we would consume fewer finite resources and design waste out of the system.

The situation has been made even more urgent by the fact that the main purchaser of our recyclable materials no longer wants to serve as the receptacle for our contaminated plastic and paper waste streams. For decades, it made sense for the U.S. to diligently collect recyclable materials and send them to China. The U.S. imports manufactured goods such as electronics, machinery, furniture, and toys from Asia in shipping containers, and those shipping containers would have returned empty if we did not fill them with something. The "something" we came up with was waste material: paper, cardboard, plastic, and metal. Over time, as China's recycling industries flourished, U.S. recycling industries languished.

Then, in February 2013, China decided there was too much contamination in the bales of recyclable materials from the U.S. China established the Green Fence policy, which set the acceptable contamination rate at 1.5% for mixed paper and mixed rigid plastics scrap, a high standard. Chinese Customs started inspecting bales at the ports, which slowed down port operations: when shippers do not load or discharge their cargo within the time agreed, the owner of a chartered ship has to pay a demurrage charge, which cuts into their profits. The inspectors of incoming cargo in Chinese ports found contamination levels in mixed paper and mixed rigid plastics scrap higher than the 1.5% acceptable contamination rate set in the Green Fence policy and thus rejected many shipments, which threw the recycling export system into disarray.[84]

From the perspective of those on the receiving end of recyclable materials, manufacturers need a clean stream of materials to make a high-quality recycled content product. This means someone must sort out the contamination found in recyclable paper or plastic bales, such as electronic waste, textiles, medical waste, insects, and food waste.

Ben Harvey, president of Massachusetts recycling company E. L. Harvey & Sons, said that through sorting, he can get contamination levels down to between 1% and 2% but no further unless someone develops better machinery or he can expand the line of workers doing manual sorting.[85] Bulk Handling Systems developed automated recycling sorting lines that move at a rate of 80 picks per minute compared with a person who can perform 30 picks per minute. Technology like this would ensure recycling clean enough for China, but the robots are expensive, and few companies have them.[86]

Then on January 1, 2018, China enacted the National Sword policy. China sought a cleaner recycling stream and lowered the acceptable contamination rate to 0.5%, a level that recycling sorting-and-baling companies in the U.S. find prohibitively expensive to meet. This left the U.S. recycling industry in a tough spot. Recycling companies started piling up bales of recyclable materials, unsure where they would be shipped. Before China's new policy went into effect, shipping ports in California were sending 15 million tons of recyclables materials to China, Korea, Taiwan, and other countries in Asia per year. Figure 20 shows the origin of recyclable materials leaving California ports.

FIGURE 20: U.S. RECYCLABLE EXPORTS
IN 2016 BY REGION (TONS)

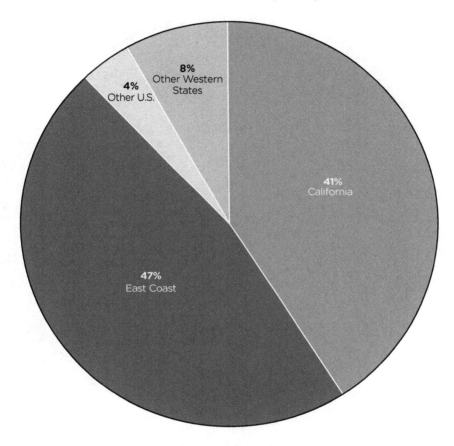

Source: CalRecycle

Figure 21 shows the types of materials shipped to recyclers in Asia in 2016, broken down by weight leaving California ports.

FIGURE 21: SEABORNE EXPORTED RECYCLABLES
BY WEIGHT (15 MILLION TONS)

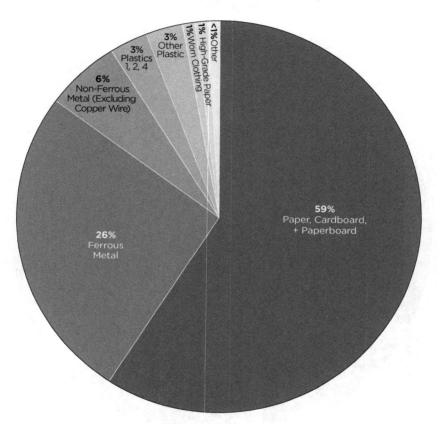

Source: CalRecycle

Given that the plastics and mixed-paper streams carry the most problematic levels of contamination and comprise two-thirds of exported recyclables, West Coast recycling brokers are, as of early 2019, scrambling to figure out what to do with these materials.

In reaction to China's tightening of standards for U.S. recyclable exports, those in the U.S. who work on recycling and waste prevention issues are contemplating the path forward. At the Northern California Recycling Association's March 2018 Recycling Update conference, Mikhail Davis, the Director of Restorative Enterprise for Interface Carpet, connected China's rejection of recyclables to the U.S.'s efforts to build a circular economy. He remarked, "Circular is great until one point in the circle stops, and then it turns linear really quickly."

Looking to the future, we could see the recycling situation as either a crisis or an opportunity to revisit how we use and dispose of materials in the U.S. Do we want to more closely examine the use of single-use materials, reduce our dependence on disposables, and switch to reusable packaging and materials? Do we want to invest in domestic infrastructure that will allow us to recycle materials into new products, rather than shipping them "away," overseas? This is our chance to overhaul our materials economy and reduce, reuse, and recycle regionally, but change needs to start at the local level.

PIVOT #13—LOCAL GOVERNMENT WASTE PREVENTION COORDINATORS

People leading the charge in the war against waste work in local government. The city staff who administer garbage and recycling hauling contracts—which they must craft carefully because many of these contracts last 20 or 25 years—know the number of tons of each material their city generates. Many cities and counties have multiple program managers who encourage the residential and non-residential sectors to recycle and compost more. By also hiring or contracting municipal waste prevention coordinators to develop and implement advisory programs for both their residential and commercial sectors, local governments would have less waste from packaging, building demolition, consumer products, discarded clothing, urban trees, and other types to manage.

The program run by StopWaste in Alameda County, California, encourages businesses to switch from single-use or limited-use transportation packaging to reusable transport packaging. One project StopWaste's "Use Reusables" program helped implement was run in conjunction with Full Belly Farm, a community-supported agriculture farm in Guinda, California. StopWaste provided a grant to help buy 2,000 reusable plastic totes to replace the waxy cardboard boxes Full Belly had been using to deliver organic produce to their community-supported agriculture drop-off locations, farmers' markets, restaurants, and grocery stores in Alameda County. The reusable plastic totes replaced

8,400 waxed cardboard boxes per year and paid for themselves in the same time period. Full Belly Farm has been thrilled with the reduced environmental impact, durability, and cost savings of the reusable totes, as have their customers.

Jobs needed:

- Waste prevention coordinators

PIVOT #14—BUILDING DECONSTRUCTION

To reduce another sizeable chunk of solid waste going to U.S. landfills, we must do more to divert construction and demolition (C&D) waste. In 2015, the U.S. produced 548 million tons of C&D debris, more than twice the amount of non-hazardous municipal solid waste. Of the total amount of construction and demolition debris created, 90% of it is demolition waste that includes:

- Concrete
- Wood (from buildings)
- Asphalt (from roads and roofing shingles)
- Gypsum (the main component of drywall)
- Metal
- Brick
- Glass
- Plastic
- Salvaged building components (doors, windows, and plumbing fixtures)
- Trees, stumps, earth, and rock from site clearing[88]

The Building Materials Reuse Association would like to see more of these materials reclaimed and reused. When deciding where to focus efforts, Joe Connell, Executive Director of BMRA, explained that houses built before the 1970s are prime targets for deconstruction: starting in the 1970s, the construction industry started gluing many building components together, which makes salvage more difficult.

Deconstructing houses built before the 1970s can provide valuable building materials that can be reused once materials are separated and nails removed. Dave Marcan with Marcan Enterprises in the San Francisco Bay Area explained that when they deconstruct a house, the most valuable materials they salvage are:

- Solid wood flooring (over 2.25" wide)
- Cabinets
- Framing wood
- Doors
- Wood sash windows
- Architectural trim
- Brick and stone

These materials can be sold or donated to a non-profit for a tax write-off, which helps cover the additional labor cost of deconstructing a building instead of demolishing it. Deconstruction specialists use a rule of thumb of approximately one week of labor for every 1,000 square feet of house; by contrast, demolition uses large, powerful equipment to smash and push over a house, load it into a roll-off dumpster, and then haul it away in about two days. The upside of demolition is a lower cost, but the downside of the process is possible dispersal of contaminants such as lead paint, asbestos, mercury, and dust throughout the neighborhood. Deconstruction is healthier for people and the environment, and it also creates six to eight more jobs for the community than standard demolition.

After pulling a building apart piece by piece, the next step in deconstruction work is making an inventory of the valuable materials to be sold or donated. Diana Pell with Re-D-Find has worked as a deconstruction project manager and currently does primarily interior design. When helping a new homeowner personalize their space, one of the first things people want to do is replace the kitchen cabinets, even if they were installed right before the new owners bought the house. "This kind of situation happens quite often," she said. Given her experience working on deconstruction projects, she knows how to find new homes for brand new but unwanted kitchen cabinets.

When Diana worked on whole-house deconstruction projects, her process for identifying salvageable building materials included:

- Creating 20 bar code labels numbered #1–#20, one for each building material item
- Entering in a spreadsheet the street address and a description for each building material, such as *36"x80" cherry door with gold peep hole and gold knob*
- Photographing each item

Deconstruction project managers then disseminate this information via their distribution lists. Ted Reiff with The ReUse People has sent such information to his 5,000-person mailing list and welcomes people to come retrieve and purchase building materials at the site. And while disseminating a list of salvageable materials from a job site to a personal list of contacts is a noble endeavor, the ability to upload this type of information to a regional online database would allow it to reach even more people.

Some organizations maintain small-scale databases but do not have the resources to keep them up to date. If each region creates and maintains a database that makes finding salvaged building materials as easy as finding things on Amazon or at Home Depot, we will make a serious dent in the amount of demolition waste that ends up in landfills. This effort deserves more financial and personnel resources in order to build a sophisticated database for each region and to update it regularly and publicize it to a wide audience.

For deconstruction to displace demolition, municipal governments must oblige deconstruction. In 2016, the City of Portland was concerned that the 300 or so older houses demolished every year were throwing away valuable building materials and thus established an ordinance requiring houses built before 1916 to be deconstructed instead of demolished. To support program development, the city created a full-time Construction Waste Specialist position to shepherd the permitting process. Shawn Wood was hired to coordinate pre-deconstruction forms, site signage, deconstruction specialist training, inspections, post-deconstruction forms, and receipts, then provide the final sign-off.

With financial support from the Oregon Department of Environmental Quality, for one year the City of Portland also provided $2,500 deconstruction grants to help close the gap between demolition and deconstruction costs and encourage establishment of deconstruction businesses. As a result, between November 2016 and October 2017, 37 buildings were deconstructed, five million pounds of materials were reused, fourteen deconstruction contractors were trained and certified, and two deconstruction contractor companies were formed. Portland has shown the steps other cities can take to replicate and scale up development of the nascent deconstruction industry and create meaningful jobs.

Jobs needed:

- Deconstruction appraisers
- Deconstruction trainers
- Deconstruction workers
- Database programmers
- Marketing specialists

PIVOT #15—TOOL LENDING LIBRARY + REPAIR CAFÉ + MAKER SPACE

Businesses doing full-time deconstruction work need their own tools, but homeowners who do infrequent construction, landscaping, and woodworking projects around the house may not necessarily need a full array. Clearly, every house needs a basic kit including a hammer, a few screwdrivers, and a power drill, but it may not make sense to buy and store a demolition hammer, tile saw, or power washer. If every household were able to borrow specialized tools from their community's tool lending library as needed, they would save money and reduce their environmental impact.

At least a few dozen tool lending libraries exist throughout the U.S. The City of Berkeley, California, started a tool lending library in 1979. The program has grown in popularity over the years to the point where they now lend out about 5,000 tools per month. The program enjoys high demand in part because the library makes sure they are well-stocked with the most popular tools in several categories: carpentry and woodworking, concrete and masonry, clamps, electrical, floor and wall, gardening and digging, ladders, material handling, mechanical tools, plumbing and drain cleaning, power tools, sanders and grinders, and saws. Demolition hammers in particular are popular, with residents checking them out 400 times per month.

Tools are loaned free to residents over eighteen years old with a driver's license and proof of residency. People can check out hand tools for three to seven days; power tools can be borrowed for three; small tools and gardening items can be borrowed for seven. Late fees are levied on items not returned on time, and popular items have a higher late fee: weed whackers and circular saw late fees are $6 per day. These late fees add up and provide a substantial revenue stream for the tool lending library, which helps pay for the full-time staff who check out and maintain tools.

If every one of the 482 towns in California had a tool lending library with two full-time workers, we could create nearly 1,000 jobs in the state alone and provide a valuable resource to each community. Berkeley's tool lending library has inspired other communities to start their own. Tool lending libraries exist all over the U.S.: Asheville (North Carolina), Columbus (Ohio), Ketchikan (Alaska), North Omaha (Nebraska), Orem (Utah), and Wichita (Kansas), among others. Each one is slightly different. The non-profit Habitat for Humanity runs the tool lending library in Santa Fe, NM. Some run by private citizens allow community members to check out tools from the property owner's barn. Some run classes, and some provide maker spaces. Others run repair cafes.

These spaces provide authentic, non-commercial places for the community to interact. Adam Broner at the Berkeley Tool Lending Library said, "My favorite part of working at the lending library is the people. I enjoy hearing about the projects they're working on and catching up on their lives."

Just like Berkeley's library has a tool lending library, public libraries may be a good location to host other services that would help the community create a circular economy and reduce waste. People want to learn how to repair broken items, creatively refashion old items, and check out tools for free. The benefits extend beyond reducing environmental impact: if every town allocated public space that hosted a tool lending library, a repair café, and do-it-yourself and maker classes, people would save money, build community, and provide kids an alternative to screen time.

Many parents agree that kids spend too much time staring at electronic screens. A greater number of public maker spaces with tools and supplies would allow kids to spend more time creating, building, and learning to repair things. By having places where people can bring broken vacuum cleaners, lamps, and toys they would otherwise throw away, children can learn from people who know how to fix things and who love to share their skills.

As more communities come to see the value of repair cafes and maker spaces, the number of facilities continues to grow. The Boulder Public Library has a U-Fix-It Clinic,[89] for example, and the Charlotte Mecklenburg Library in North Carolina has a maker space called the "Idea Box" where youth can learn 3D printing, knitting, and coding.[90]

Libraries that want to start their own Remakery with a tool lending library, repair café, and maker space could pilot one on a small scale and grow from there. Nick Szegda, Associate Director for the Menlo Park Library in Menlo Park, California, suggests starting this process by answering a few questions:

- What are several tools that people in the community will likely want to borrow?
- What are several common items people would want to be able to fix but don't know how?
- What are a few things people might want to be able to build?

Answers to these questions will provide the building blocks for a business plan. Szegda suggested requesting donations for tools through community forums like NextDoor.com. Next, the library will need a locking cabinet, whether donated or purchased, to store tools for lending. Then the program manager would fundraise for the rest of the needed budget.

When developing classes for a maker space that teaches people how to build new things from old and new materials, think about what the public might be interested in building. Classes that teach people how to turn a standard bicycle into an electric-assist bicycle have been popular, Szegda explained. As well, classes that help children build things with batteries, lights, and motors are popular. The skills learned in these spaces will help society move toward a circular economy in which we repair and upcycle items instead of buying new items.

Jobs needed:

- Tool lending library site coordinators
- Project managers
- Instructors

PIVOT #16—UPCYCLING DEAD OR DISEASED TREES

Every year, removed or fallen trees in U.S. urban areas produce the equivalent of 3.8 billion board feet of potential lumber, which is more than 25% greater than our entire national forest system's output.[91] Yet, when a building owner contracts with a tree removal company to extract a tree from their property, that tree is often reduced to piles of firewood, mulch, and sawdust, then hauled away.

There's a better way. We can upcycle urban trees into lumber to be used in construction and furniture-making.

Portland-based urban lumber advocate Dave Barmon explained how to safely take down a tree in a front yard so that the base can be turned into lumber (keep in mind that every situation is different). The basic process involves an arborist climbing the tree with rigging, gear, and a chainsaw. First, they remove all tree limbs, reducing the tree to the bottom ten to twenty feet. If there is room available, the bottom stem of the tree can be dropped in lengths long enough to produce saw logs. (Generally, saw logs should be over eight feet in length.) Next, they stop to consider how best to fell the trunk of the tree: in which direction could they safely bring down a ten-, twenty-, or thirty-foot length of trunk so it will not hit anything? They then guide the base as they make the cut and drop the log to the ground.

Once the logs are on the ground, they can be staged for pick up with heavy equipment or a peavey can be used to roll smaller logs. A portable sawmill can be brought to the site to mill logs into lumber on the spot. Alternatively, someone with a grapple truck containing a hydraulic arm will need to pick up the log, place it on the truck bed, and haul it to the nearest sawmill.

Urban Hardwoods in Seattle, Washington, intervenes early in the process to purchase logs directly from property owners. Depending on the tree size, height, and species, Urban Hardwoods may pay $300–$500 for the logs from a removal. They mill and dry the wood themselves, then manufacture it into furniture like dining tables, conference tables, desks, benches, stools, and bed frames.

One of Urban Hardwoods' favorite pieces of furniture to make is a "forever table." Big, old trees where children have climbed and played become a beloved part of a family's property. When such a tree becomes diseased or dies and needs to be removed, turning it into a "forever table" allows the tree to live on as a cherished piece of furniture for the family.

Dave Hunzicker, Operations Manager at Urban Hardwoods, said that with the recent uptick in hurricanes, he's thought about a side business for upcycling the large number of previously healthy trees uprooted during hurricanes. Typically hauled to landfills, he would rather see these trees turned into lumber for reconstruction efforts in hurricane-damaged areas.

Hunzicker explained that aspiring entrepreneurs wanting to start up a tree salvage and log milling business would need the following equipment to extract toppled trees and cut them into boards at the site.

- **Chainsaws**—essential equipment for cutting logs to the right length
- **Portable sawmill**—a trailer-mounted sawmill that attaches to the back of a pick-up truck and is driven to the site to enable processing of two to three logs per hour (cost: about $30,000)
- **Chainsaw mill**—works much slower than a portable sawmill but can be carried by two people and enables processing of two to three logs per day (cost: $4,000–$5,000)
- **Chains, straps, and cant hooks**—to move logs around on-site and into the mill
- **Pickup truck** with a trailer or flatbed truck

This presents a solid business opportunity in places in the Southern and Eastern U.S. where hurricanes and tropical storms hit each year between June and November. This also provides an opportunity in areas out West where drought has killed hundreds of millions of trees. Instead of cutting down healthy trees in national forests and plantation forests, we can source our furniture and construction lumber from trees whose demise was precipitated by natural disasters or prolonged drought.

Jobs needed:

- Tree removal specialists
- Drivers
- Sawmill operators
- Furniture makers

PIVOT #17—ARTISTIC UPCYCLING AND SALVAGE

Old clothing is another example of a waste item that could be upcycled into fun new products. All we need are creative minds to devise and execute exciting new uses for them. Clothing company Elisabethan sources old T-shirts from thrift stores, cuts them up into specific shapes, and sews them into new shirts and skirts. Twelve old T-shirts will become one new shirt in a funky, innovative design. In our "fast fashion" society, where people buy and wear some clothing for just one season, Elisabethan extends the useful life of textiles.

Some thrift stores have been hiring fashion designers to figure out ways to upcycle and sell more of the donations they receive. St. Vincent de Paul of Lane County, Oregon, hired fashion designer Mitra Chester full time to help increase revenues. Chester started by working on the sorting line to gather a better understanding of the donations coming to the store. She saw Western and military-style clothing items being rejected from the donation piles but thought students at the nearby University of Oregon might like these items, so she fished them out and displayed them in the store. They sold quickly.

Chester then started looking for opportunities to upcycle worn-out, torn, stained, or partially used clothing into fresh and exciting items. She cut sleeves off flannel shirts and hoodies, then sewed the flannel sleeves onto the hoodies' bodies. The first batch sold out in short order, so she kept making them. She cut up old belts and studded them to make bracelets. She cut the legs off old blue jeans to make cutoffs, then sewed the legs into purses. She cut off the tops of cowboy boots and made purses out of those, too. She contracted with a company that does laser cutting to convert scratched vinyl records into earrings. Old ties were fashioned into cell phone holders. And for those whose purses can never hold enough stuff, Mitra cut out leather pieces from old, ripped, and stained couches and sewed them into large shoulder bags.

To expand her marketplace beyond Lane County, Chester sells her creations online at enviafashion.com. The revenue goes back to non-profit St. Vincent de Paul to help them meet their mission of providing emergency services to the homeless and low-income communities. With a $42 million annual budget, St. Vincent de Paul provides five shelter programs, 1,400 units of affordable housing, and emergency assistance for people who recently lost their job or home. As the majority of St. Vincent de Paul's revenue comes from waste diversion-based social enterprises, the more money they can raise, the more they can help the local community.

Hiring a fashion designer to upcycle store donations was a smart move on the thrift store's part. Mitra's creativity has helped increase revenue in her store from $500 to $1,500 per day. Susan Palmer, Economic Development Director of St. Vincent de Paul, explained that in places like thrift stores, with a large inflow of material donations and a ready audience to whom they can market their creations, bringing on a fashion designer provides new revenue streams and facilitates a higher and more creative use of materials. More thrift stores should hire fashion designers to replicate Mitra Chester's success and divert more of the river of clothing waste heading toward landfills every year, as well as increase the revenue that supports their mission.

Another place where clothing goes to waste is in stores. Many retail items cannot be sold because the shipping box was soaked by rain or an item lost a button, has a broken zipper, or suffered a makeup stain in the fitting room. Higher-end stores like Nordstrom boast on-site resources to repair and clean garments so they can freshen them up and put them back onto the show floor, while other retailers need outside help to do this.

The Renewal Workshop in Cascade Locks, Oregon, works with brands like prAna, Ibex, Toad&Co, Mountain Khakis, Indigenous, and Thread by taking their lightly damaged clothing items and fixing or cleaning them. Jeff Denby, cofounder of The Renewal Workshop, describes a "tsunami of rejected clothing moving back toward the brands." He explained that the brands usually shred rejected clothing and send it to the landfill or "debrand" it by stripping out the label and then donating it to charity. If a greater portion of these millions of tons of clothing landfilled each year are cleaned, fixed, and sold—even at a discount—we will reduce the environmental impact of the fashion industry.

In 2017, The Renewal Workshop saved 42,000 pounds of apparel from landfill. Clearly, there exists an opportunity to divert much more clothing from the waste stream: worldwide, the clothing industry dumps 10.5 million tons of clothing in the landfill every year.

A 2017 study from the Ellen MacArthur Foundation[92] investigated ways to reduce waste in the textile economy and recommends four measures.

FIGURE 22: CREATING A NEW TEXTILES ECONOMY

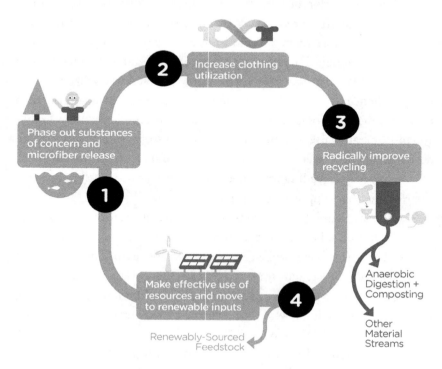

Source: Ellen MacArthur Foundation

After phasing out substances of concern and minimizing microfiber releases, we will, as a society, find ways to increase clothing durability. This involves making clothing that lasts longer than the one-season fast fashion widely available, which features fabric or seams that can only handle several washings.

Throwing clothing into the landfill creates few jobs, but by reducing the environmental impact of the clothing industry, numerous opportunities can be created. Alternatively, we can create jobs to freshen up damaged or stained pre-consumer clothing, upcycle used clothing into something new, or even downcycle blue jeans into building insulation. We will also need marketing and social media experts to let the public know about these new creations.

Jobs needed:

- Fashion designers
- People who sew
- Photographers
- Online and social media marketing specialists

PIVOT #18—REGIONAL RECYCLING MARKET DEVELOPMENT MANAGERS

For further inspiration about how to reduce the environmental impact of our materials economy, we should look to nature. Janine Benyus's seminal book *Biomimicry* describes how nature "manufactures" by following some basic guidelines. Nature "shops" locally for raw materials by using materials found nearby. Nature also uses materials sparingly: consider a bee colony's honeycomb with hexagonal cells that share wax walls. In nature, there is no waste, as waste materials from one process become feedstock for another.

Contrast nature's approach with the 262.4 million tons of municipal solid waste generated in the U.S. in 2015. Of those 262.4 million tons, the U.S. recycled 68 million and composted 23 million.[93] Not only is landfilling valuable materials a waste of resources, but recycling and composting employ more people per 1,000 tons than disposal does. Creating more recycling and composting jobs would help put people with entry-level skills to work building a circular economy.

A 2008 Tellus Institute study calculated how many jobs we will create if the U.S. establishes a National Strategy for achieving 75% waste diversion by 2030.[94] Using baseline data from 2008 of 250 million tons of municipal solid waste, the Tellus Institute report looks at five materials that comprise 77% of the total municipal solid waste generated in 2008: paper and paperboard, yard waste, food scraps, plastics, and metals. On top of the 250 million tons of solid waste, the U.S. generated 178 million tons of construction and demolition debris in 2008. Seventy percent of this total was concrete, mixed rubble, and wood. Figure 23 shows the number of MSW and C&D waste handling jobs that existed in 2008 (with an average 33% recycling rate), the number of jobs projected to be created if current trends continue through 2030 (called the 2030 Base Case), and the number of jobs we will create in a Green Economy Scenario if we set a 75% landfill diversion goal.

FIGURE 23: TOTAL MUNICIPAL SOLID WASTE AND CONSTRUCTION & DEMOLITION JOBS IN U.S.

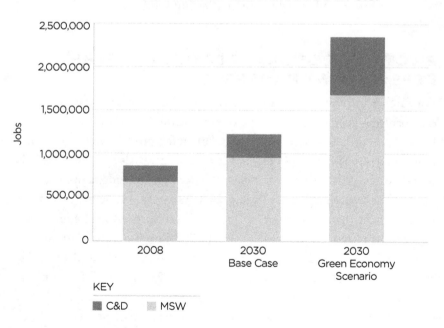

Source: U.S. Environmental Protection Agency

By diverting 75% of waste from landfills, Tellus calculates that we could almost triple the number of jobs to over 2.35 million by 2030. This is partly because waste disposal is more equipment-intensive, not labor-intensive, and for various waste management activities generates only 0.1 jobs per 1,000 tons. By contrast, recyclables processing creates two jobs per 1,000 tons, and composting creates 0.5 jobs per 1,000 tons. Once recyclers reclaim these materials, manufacturing of these materials into new products creates additional jobs: four jobs per 1,000 tons for paper, iron, and steel and 10 jobs per 1,000 tons for plastics.[95]

Overall, the Green Economy Scenario in the Tellus study projects that setting a 75% diversion rate throughout the U.S. by 2030 will create 2,347,000 jobs, 1.1 million more jobs than in the 2030 Base Case and nearly 1.5 million jobs more than in 2008.

Many cities that have adhered to aggressive waste diversion goals for decades and have devoted resources to creating strong recycling and composting already enjoy high diversion rates. In 2017, Fremont, California, a city with a population of 234,000 residents, achieved a 67% waste diversion rate. To consider how many jobs could be created if Fremont's waste materials were recycled in the region, let's look at their waste characterization. In 2016, Fremont's residential sector generated 58,057 tons of waste[96] and its commercial sector generated 149,633 tons of waste. Table 3 (on the next page) shows the breakdown of waste materials generated that are recyclable.

Given the Tellus Institute's assumptions that landfilling creates 0.1 jobs per 1,000 tons, compared to two jobs for every 1,000 tons recycled—and on top of that four jobs for every 1,000 tons of paper recycled and ten jobs for every 1,000 tons of plastic recycled—Fremont already creates many more jobs through recycling and composting than if its waste were landfilled.

Now, if Northern California or nearby Central Valley were to develop recycling industries to take Fremont's waste materials and turn them into new products, we would create jobs for people in the region. Considering that in 2016, Fremont's residential and commercial sectors generated 1,198 tons of PETE (#1) plastic, 794 tons of HDPE (#2) plastic, 19,062 tons of cardboard, and 2,765 tons of white ledger paper, sorting and remanufacturing these materials in the region would create 154 jobs versus two jobs sending waste to landfill.

TABLE 3: SELECTED MATERIALS IN 2016 WASTE STREAM—
CITY OF FREMONT, CALIFORNIA

CATEGORY	RECYCLABLE MATERIAL	TONS FROM RESIDENTIAL SECTOR	TONS FROM COMMERCIAL MULTI-FAMILY SECTORS
Plastic	PETE #1: Plastic	383	815
	HDPE #2 Plastic	250	544
Paper	Uncoated Corrugated Card-board	787	18,275
		211	472
	Paper Bags	959	2,904
	Newspaper	155	2,610
	White Ledger Paper	293	2,481
	Other Office Paper	382	1,218
	Magazines and Catalogs	19	50
	Phone Books and Directories	1,929	5, 969
	Other Miscellaneous Paper		
	Other Miscellaneous Paper—Compostable	103	N/A
	Remainder/Composite Paper—Compostable	5,743	N/A
	Remainder/Composite Paper—Other	543	N/A

Source: CalRecycle

For the past 20 to 25 years, recycling industries in the U.S. have atrophied as more and more recyclable materials were sent to China and other points in Asia. Now that China is rejecting more and more loads of recyclables, some materials are heading for Indonesia, India, Malaysia, and Vietnam. Buyers in these countries pay $6 per ton for clean cardboard, and there is essentially no market for plastics marked #3 through #7 (the numbers inside the chasing-arrows symbol on the bottom of plastic containers).

Development of more recycling industries in the U.S. would face some headwinds but none that are insurmountable. New recycling facilities will require capital to site, build, permit, and staff them. Creating facilities that meet environmental regulatory standards and addressing Not In My Back Yard (a.k.a. NIMBY) resistance from potential neighbors will take extra effort on the businesses' part.

Regional Recycling Market Development Managers could help navigate these types of challenges as they work to build regional recycling markets. Part of this work involves helping large businesses who want to invest in and open a new facility in the region: the Recycling Market Development Manager would help identify a site, low-interest loans, tax credits and grants and assist with permitting. This regional manager could also help entrepreneurs who want to launch small businesses that recycle paper and plastic materials into new in-demand products. Ecologic in Manteca, California, is one such business.

Founded in 2008, Ecologic manufactures paper bottles made from recycled materials. These bottles come in different shapes for holding wine and spirits, food and powder, and pet care, home care, and personal care products. The molded-fiber outer shell, which is made from 70% recycled corrugated cardboard and 30% newspaper, can be composted. The inner liner is made from either #2 plastic, #4 plastic, or bioplastic.

For forward-thinking cities, counties, and states that want to develop industries in their region, there exist a host of products that can be manufactured from recyclable paper and plastic materials collected from residential and commercial sectors. Recyclable paper could become paper water bottles and modular furniture such as storage cubes, shelves, tables, and desks. Recyclable plastics could be remanufactured into reusable transportation packaging (such as totes, pallets, and pallet wrap), reusable mail-order packaging for two-day shipping, and composite lumber for non-structural applications. The following are examples of each of these options.

Paper Water Bottles Can Eliminate a Massive Waste Item

Similar to Ecologic, the company Paper Water Bottle, based in Walton, Kentucky, offers a packaging product made from biodegradable and recycled content. The bottle's outer layer and cap are made from biodegradable bamboo, bulrushes, and sugar cane, while the inner bladder contains a 100% recycled resin barrier.

The idea for Paper Water Bottle came to inventor Jim Warner, at the time a consumer packaged goods designer, while walking along a busy city sidewalk with his young son. His son asked, "Dad, what do you do for work?" Warner saw a bottle lying in the gutter, pointed to it, and said, "I design and make packaging, like that bottle there." Confused, his son replied, "You make trash?" This partly inspired Warner to want to make more environmentally friendly packaging, and the idea for a new company was born.[98]

Modular Furniture Makes for Easy Redesign of Interiors

Several companies take waste paper and cardboard and recycle them into pressed squares for modular furniture. Adhesive strips along the edges allow the squares to be assembled into storage cubes, shelves, tables, or desks without using nails or screws. The modular design means people can reassemble the pieces in different configurations to periodically update their interior designs. These products use recycled content paper pressed together with non-formaldehyde adhesives.

Reusable Transportation Packaging Prevents Waste and Is Easier on Workers

Standard secondary packaging, meaning transportation packaging that protects the primary packaging enveloping a product, often consists of cardboard boxes stacked on wooden pallets sealed with stretch wrap. This single-use or limited-use packaging is convenient but creates large amounts of waste, as cardboard is often recycled or thrown away after only one use. Wooden pallets are often used for three to five trips before needing to be rebuilt or are discarded. Stretch wrap is rarely recycled.

A more durable alternative transport packaging option would be reusable totes stacked on reusable pallets wrapped with reusable pallet wrap. All three items could use recycled plastic material for a portion of the feedstock. One company that uses all three reusable transport packaging items is Veritable Vegetable, which distributes produce grown in the Central Valley to businesses throughout San Francisco. A few years ago, they switched to reusable totes, pallets, and pallet wrap. With an initial investment of $75,000, they now save $46,000 per year by not having to buy cardboard boxes, wooden pallets, and plastic stretch wrap.

The company also realized an important ergonomic benefit from the switch to reusables. Warehouse shipping staff previously packaged up produce by loading cardboard boxes onto wooden pallets, then walked around each pallet several times with a roll of stretch wrap while hunched over in order to secure the load. Switching to reusable pallet wrap involves less bending: in the new, reusable wrapping system, warehouse staff simply need to walk around the pallet once to encase the load with a wrap, then thread three Velcro straps through three sets of D-rings before tightening—and they are done securing the pallet. This more ergonomically and environmentally beneficial transportation packaging saves the company staff time and money.

While a Local Government Waste Prevention Coordinator can help a company prevent waste by switching to reusable transport packaging, a Regional Recycling Market Development Manager can help develop companies that can manufacture the reusable transport packaging out of regionally sourced #1 and #2 plastics. Both roles are important. Setting up reverse logistics systems for recycled-content transport packaging will also create demand for materials, which will be collected regionally.

Reverse logistics systems for packaging mailed to households can also prevent waste while using recycled-content materials in reusable transport packaging. Currently, people who order products online receive the items in a cardboard box. With the growth of online shipping, the number of mailing boxes used in the U.S. is increasing.

One company, Toad&Co, is experimenting with a reusable mailer by shipping their clothing to customers in a durable vinyl pouch made from old billboard signs. Their mailers come from Limeloop, that makes a lightweight, waterproof, zippered pouch that can be reused up to 2,000 times.

A mailing label slides into a clear plastic window, which can be pulled out once the package reaches its destination. Customers then pop the reusable mailers back in the mail to return them to Toad&Co for reuse—all for free, as long as they use the code from the back of their invoice.[99]

Yesterday's Fishing Nets Become Tomorrow's New Textiles

Finally, in the interest of creating a market for a large portion of the plastics floating in the Great Pacific Garbage Patch,[100] Italian yarn producer Aquafil is spearheading an effort to recycle lost and abandoned fishing nets. Fishing nets are often made from nylon 6 plastic, and they comprise 46% of the plastic in the Great Pacific Garbage Patch. Each year, 705,000 tons of nylon 6 fishing nets go missing around the world and end up ensnaring 136,000 seals, sea lions, and whales. These missing fishing nets are referred to as "ghost gear."

The Healthy Seas Initiative, a joint venture of non-governmental organizations and businesses, retrieves and recycles these abandoned fishing nets with the help of volunteer scuba divers and fishermen. Aquafil pays fishermen in harbors to strip away the seines and gillnets to prepare old nylon nets for shipping. The nets are then packed into bags that, when full, weigh 800–1,000 pounds; once 40,000 pounds of nylon have been packed, the bags are shipped to Aquafil's recycling facility in Slovenia. Between 2013 and 2017, Aquafil sent hundreds of tons of fishing nets to the plant where they were shredded into fluff, then shipped to a depolymerization plant and turned into nylon pellets for use in manufacturing new socks, swimwear, carpets, and other textiles.

We must ramp up this effort to reclaim lost and abandoned nylon 6 fishing nets and establish recycling plants along the U.S. coasts. Doing so will create deeply meaningful jobs while removing plastics from the oceans.

Jobs needed:

- Regional recycling market development managers
- Entrepreneurs
- Finance specialists
- Marketing specialists

CHAPTER HIGHLIGHTS

- The U.S. has been sending recyclable materials to China and other countries in Asia for decades. Over that time, demand for recyclable materials in the U.S. withered, as did regional recycling infrastructure that could take and use recyclable materials.
- Moving from a take-make-waste linear economy to a circular economy will involve investment in regional recycling and composting infrastructure, reuse projects, and waste prevention projects.
- Many jobs in a circular economy can employ people with entry-level skills, such as those recently released from prison.

7

MEANINGFUL JOBS FOR

REDUCING FOOD WASTE

"To eat responsibly is to understand and enact,
so far as one can, this complex relationship."
—Wendell Berry, poet and farmer

When 400-acre Full Belly Farm in Guinda, California, harvests oddly shaped or bruised produce they cannot sell, nothing is wasted. They put it to good use by feeding it to their 200 sheep and 1,000 chickens, thus turning it into a value-added product. Unsellable produce provides sustenance to sheep that produce wool, baby lambs, and meat and to chickens that produce eggs and meat. Or, the farm's kitchen turns surpluses of delicate, easily bruised items into tasty jarred items with a long shelf life. Extra tomatoes become pasta sauce or sun-dried tomatoes. Strawberries are turned into jam. Basil is made into pesto.

Creating value-added food items yield a higher financial return for the farm, but composting some of the misshapen or bruised produce and applying it onto the soil is also important. Compost feeds microbes in the soil, which improves soil health and results in better crop yields.

Those who visit Full Belly Farm during their annual Hoes Down Festival in early October will see how dark and humus-rich the soil is.

One of the farm's owners, Judith Redmond, has a master's degree in Soil Science from the University of California, Davis and decades of experience growing organic produce. She knows that managing the balance of nutrients, microbes, and moisture in the soil by feeding organic waste back into it as compost is key to growing healthy organic produce for farmers' markets, grocery stores, and restaurants in the region. Waste from one process becomes food for another.

Large monoculture farms and confined animal feeding operations (CAFOs) have a different approach to agriculture, keeping animals and crops separate. These systems drive down prices but also set the stage for waste throughout the food supply chain. In 2016, when 49 million Americans struggled to put food on the table, the U.S. threw away 52.4 million tons of food.[102] By creating jobs that prevent, redirect, or compost food waste, we will reduce hunger, create more small businesses and non-profits, and reduce greenhouse gas emissions.

According to Paul Hawken's *Drawdown*, reducing food waste sits at #3 on the list of projects with the greatest potential to reduce atmospheric greenhouse gases. *Drawdown* calculates that between 2020 and 2050, reducing food waste around the globe has the potential to reduce atmospheric greenhouse gases that contribute to climate change by 70.53 gigatons. The misallocation of food not only denies nourishment to people; food rotting in landfills also contributes to climate change by converting into methane, a powerful greenhouse gas. Diverting more edible food, either raw ingredients or surplus prepared food, to those who need it reduces methane.

Americans are a generous people. Every year, 80 to 100 million citizens donate their money, time, and food to help fight hunger. In the U.S. in 2016, approximately 200 food banks and 60,000 food pantries and soup kitchens served four billion meals to 46 million people.[103] These donations help the one-in-six Americans who suffer from food insecurity, which the U.S. Department of Agriculture defines as "lack of access, at times, to enough food for all household members." Food insecurity numbers are worse for children (one in five children are food insecure) and even worse for minorities (one in three African-American and Latino children at times go without sufficient food).

Unfortunately, a large amount of the food donated to homeless shelters and food banks has low nutritional value. One representative from a food recovery organization said that they receive 17,000 pounds of baked goods each month, which is 7% of the donations they receive or 20% by volume. Forty percent of those baked goods are sugar-filled pastries.[104] The emergency food system receives more sheet cakes, cookies, and pastries than their clients need with the spillover often going to animal feed. Andy Fisher, author of *Big Hunger*, reports that one pig farmer in Wisconsin started accepting donated pastries to feed his pigs, but after feasting on pastries for a month, the pigs became so aggressive that the farmer was forced to discontinue feeding them the sugar-filled desserts.

To provide more nutritionally balanced meals to people who are food insecure, the emergency food system needs more healthy food. Since food waste occurs throughout the supply chain, there are opportunities to divert it from farms, distribution centers, grocery stores, restaurants, hotels, caterers, and the residential sector.

According to ReFed's 2016 report "Roadmap to Reduce U.S. Food Waste 20%," the majority of food waste happens at consumer-facing businesses and in homes. Figure 24 shows the breakdown of the $218 billion of food that is wasted.

FIGURE 24: FOOD WASTED BY WEIGHT—63 MILLION TONS

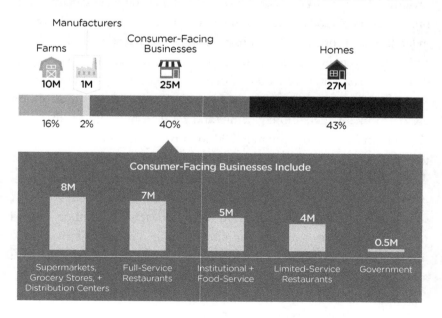

Source: ReFed

Drilling down on the retail and consumer levels, Figure 25 shows what types of food items are wasted most often.

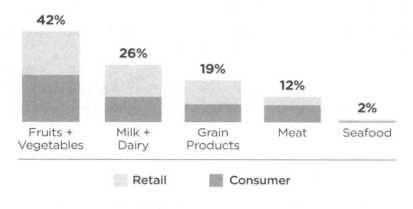

FIGURE 25: FOOD WASTE BREAKDOWN (BY WEIGHT AND TYPE)

Source: ReFed

When we add up the dollar value of food wasted throughout the supply chain, the numbers are staggeringly high. The U.S. spends $218 billion per year, which is equal to spending 1.3% of our gross domestic product on growing, transporting, and disposing of food that's never eaten. Each year, 52.4 million tons of food is sent to landfills, and an additional 10.1 million tons remains unharvested at farms, totaling roughly 63 million tons of annual waste.[105]

JOBS TO PREVENT, RECOVER, OR RECYCLE FOOD WASTE

The landmark ReFed report provides 27 solutions to prevent, recover, or recycle food waste and to create 15,000 permanent jobs. Eighty percent of the jobs would come from growth in a centralized compost sector: constructing, managing, collecting, and processing food waste.[106]

At the retail level, food is worth roughly $2.50 per pound, magnitudes higher than the value of food scraps for disposal. As we figure out where to allocate resources to prevent and divert food waste, we should consider projects with higher potential savings. The following is a list in descending order of economic value of food waste prevention and diversion projects.

$4,000-$5,000 Per Ton

Standardized date labeling—Standardizing food label dates such as "sell by" dates, which confuse consumers

Consumer education campaigns—Large-scale consumer advocacy campaigns to raise awareness of food waste and education consumers about ways to reduce food waste and save money

$3,000-$4,000 Per Ton

Packaging adjustments—Optimize food packaging size and design to ensure complete consumption by consumers and avoid residual container waste

$2,000-$3,000 Per Ton

Donation matching software—Connects individual food donors with potential recipient organizations for smaller-scale food donations

Donation liability education—Educating potential food donors on donation liability laws such as the Good Samaritan Law

Value-added processing—Extending usable life of donated foods through processing methods such as making soups, sauces, or other value-added products

Donation storage and handling—Expanding temperature-controlled food distribution infrastructure and labor availability to handle additional donation volumes

Donation transportation—Providing small-scale transportation infrastructure for local recovery as well as long-haul transport capabilities

Waste tracking and analytics—Providing restaurants and prepared-food providers with data on wasteful practices to inform behavioral and operational changes

Trayless dining—Eliminating tray dining in all-you-can-eat restaurants to reduce food waste

Smaller plates—Providing consumers with smaller plates in self-serve, all-you-can-eat dining settings to reduce consumer waste

Cold chain management—Reducing product loss during shipment to retail distribution centers by using direct shipments and cold-chain-certified carriers

$1,000–$2,000 Per Ton

Manufacturing line optimization—Identifying opportunities to reduce food waste from manufacturing/processing operations and product line changeovers

Donation tax incentives—Expanding federal tax benefits for food donations to all businesses and simplifying donation reporting for tax deductions

Improved inventory management—Improvements in retail inventory management systems' ability to track an average product's remaining shelf-life and reduce how long an item goes unsold

Produce specifications—Integrating the sale of "ugly produce" for use in food service and restaurant preparation and for retail sale

$0-$1,000 Per Ton

Secondary resellers—Businesses that purchase unwanted processed food and produce directly from manufacturers and distributors for discounted retail sale to consumers

Home composting—Keeping a small bin or pile for on-site waste at residential buildings

Centralized composting—Transporting waste to a centralized facility where it decomposes into compost

Of all these solutions, prevention and recovery solutions at the top of the list are the most cost-effective. By contrast, the composting solutions at the lower end of the cost-effectiveness list are the most scalable. Now, where specifically do we want to create jobs to do this work? The following are proofs of concept for four of these nineteen solutions. Keep in mind that this is not an exhaustive list, just evidence that jobs like these exist and deserve to be replicated and scaled.

PIVOT #19—REVERSE CATERING

When Salesforce, Dropbox, and AirBNB are left with at least ten trays of surplus catered food from events or employee lunches, they log onto RePlate's app and order a pick-up. They know that RePlate, a non-profit, will quickly dispatch a driver to collect the surplus meals and deliver them to an organization in the area that serves the homeless or food-insecure.

RePlate helps companies track their food donations' impact. This data is used by clients for Corporate Social Responsibility reports. Through RePlate's app, they can learn about the number of meals they donated, the pounds of food redirected, the gallons of water saved, and the amount of greenhouse gases prevented from entering the atmosphere, all based on data reported by pick-up drivers.

For this reverse catering service, RePlate charges $40 per pick-up or $120 for four pick-ups per month. Companies that pay for this service feel $120 is a small price for the benefits they receive: prompt pick-ups and enhanced employee engagement, the latter of which many companies struggle with. It turns out, however, that donating excess food to people who need it is a good way to boost employee pride in their organization.

When RePlate started out, they thought companies would value the financial benefits, such as a tax credit, from donating food to a non-profit or the avoided cost of disposal from not throwing away food. The factor that motivated people most to pay for pick-ups though, was that it was the right thing to do.

For many people, the main obstacle to donating surplus prepared food is the widespread misconception about potential liability. Many mistakenly believe that if they donate food and someone who eats it becomes sick, they could be sued. Fortunately, the Good Samaritan Law, a federal law that has been on the books since the 1990s, states that if a business donates surplus food in good faith and someone becomes sick as a result, they will not be held liable. If more people knew about the federal Good Samaritan Law, which has been duplicated in many states, they would feel more comfortable donating food without fear of liability.

Founder Maen Mahfoud expanded RePlate's operations into 300 cities in the U.S. and Canada in the last two years but has a counterintuitive end-goal in mind: he hopes that, over time, companies will generate less surplus food that needs to be picked up. "The best thing we can do is put ourselves out of business. This probably is not going to happen in my generation though."

Jobs needed:

- Logistics coordination specialists
- Drivers

PIVOT #20—COMMUNITY KITCHENS

"We need to shift the perspective of food recovery from charity to it being viewed as a valuable service," said Dana Frasz, Executive Director of Food Shift in Oakland, California. One service Food Shift provides to businesses with a large throughput of food is finding ways to divert edible food from landfill or compost streams to non-profits. The savings can be sizeable. After working with Food Shift, grocery store Andronico's saved $27,000 per year by avoiding garbage disposal and compost pick-up costs, as well as tax deductions for donations and savings on regulatory compliance costs. A further benefit at Andronico's was that 87% of their employees reported that the food donation program increased their sense of pride and joy in their workplace.

Frasz is pushing for companies with excess food to pay for the service. She explained that for every few thousand dollars per month Food Shift collects for food recovery efforts, they can create one full-time job for someone who is trying to become self-sufficient at the Food Shift Kitchen. "Why not chip in to make that possible?"

Food Shift Kitchen at the Alameda Point Collaborative is doing more than just picking up surplus food: they are also training people for culinary careers. Alameda Point Collaborative helps people who are homeless and struggling to get back on their feet with job training. The majority of people at Alameda Point Collaborative have a disability, are survivors of domestic violence, or are veterans. Residents feel isolated and stuck socioeconomically. Only 10–15% are employed.

In service of helping this population of enthusiastic potential workers who want to pull themselves up economically, the Food Shift Kitchen offers six-month apprenticeships, during which apprentices receive on-the-job training for culinary employment while earning minimum wage. Such food preparation and catering work at the Food Shift Kitchen provides a bridge to the workforce. After the apprenticeship, people graduate to kitchen assistant or driver, which pays close to $20 per hour. This kind of work allows people to develop skills that will help them lift themselves out of poverty.

Jobs needed:

- Project managers
- Training experts

PIVOT #21—UGLY PRODUCE DISTRIBUTION

Imperfect Produce in Emeryville, California, applies clever marketing backed by private equity funding to the problem of food waste. Their focus is on the lumpy, oddly shaped fruits and vegetables that are too ugly to be sold: carrots with two legs or cucumbers bent in a semi-circle. According to Imperfect Produce, one in five fruits and vegetables grown do not meet conventional aesthetic standards because they are:

- Too small, too big, or too varied in size
- Too misshapen
- Too marked on the peel
- Too unknown—broccoli leaves and stalks
- Too different from typical color
- Too numerous

The company's marketing genius relies on the ability to highlight and celebrate the beauty in the odd. After all, a heart-shaped potato will still taste good after it's roasted with olive oil, garlic, rosemary, and salt.

Sometimes, aesthetics aren't even the reason produce goes to waste though. Imperfect Produce rescued a shipment of Meyer's Lemons that had just travelled 8,000 miles from New Zealand and was about to be thrown away, all because the local Meyer's lemon crop in California was ready early. As a result, there was no market for the New Zealand shipment. Imperfect Produce saved the shipment from waste disposal and sold the lemons.

According to Imperfect Produce buyer Megan Langner, they work with medium-sized growers and shippers who often have to make difficult decisions about what to do with their ugly produce. Should they sell it to processors that will turn it into frozen foods, soups, packaged salad mixes, and other items, which barely covers their costs? If the market is flooded with excess off-grade items and processors are paying less or stopped buying it altogether, what should they do with it? Excess strawberries in-season can be one example: should they send the surplus product for animal feed, which pays very little? Should they plow it under, in which case the energy and water embodied in the food will be wasted? Or should they compost it, which will cost money? Growers, shippers, and distributors face decisions like these regularly, and often the cheapest, easiest solution is to simply throw it away.[107]

If, instead, farmers and shippers could send produce to Imperfect, doing so would get it into the hands of consumers, which is a higher dollar value option. Since 2015, Imperfect has recovered twenty million pounds of ugly produce. The company now has nearly 500 employees and distributes 700,000 pounds of produce each week. Given that farmers generally throw away 20% of their produce, this means millions more in the pockets of farmers. Imperfect also takes care of lower income folks with a 33% discount and donates 20,000 pounds of product to food banks.

As they look to scale up further, Imperfect will need more logistics, quality assurance, marketing, and outreach specialists. They're adding routes in cities they currently serve and expanding into new cities. By 2024–2025, they plan to recover one billion of the six billion pounds of food waste on farms. Longer term, they plan to rescue more value-added products like peaberry coffee, which are coffee beans that are too small to go to market, and packaged food like Clif Bars with freshness dates less than 60 days away.

Sometimes, the solution to a big problem is as simple as clever marketing: take the unloved lumpy, undersized, or off-color produce items and highlight their quirky side. Just as it's fashionable to adopt rescued dogs and cats, it's now cool to rescue abandoned produce.

Jobs needed:

- Logistics specialists
- Quality assurance specialists
- Marketing specialists
- Outreach and education specialists

PIVOT #22—BUSINESS SERVICES SUPPORTING SMALL ORGANIC FARMS

Imperfect Produce works with medium-sized growers to create a market for their surplus produce. Meanwhile, small farms like Full Belly Farm do not generate food waste because they incorporate unsellable produce back into their farm processes. To further reduce food waste in our society, one thing we should do is nurture the next generation of small, organic farmers.

Full Belly does its part by running a year-long internship program. Each year, they host five paid interns who learn about soil science, marketing, distribution, and how to make sure nothing goes to waste. Over the past 30 years, Full Belly has managed 200 interns, many of whom go on to start their own organic farms.

Non-profit Kitchen Table Advisors (KTA) also helps nurture a number of successful small, organic farms and ranches in Northern California by providing business planning and financial management assistance. Founder Anthony Chang, was inspired to start the non-profit the day he visited his local farmers' market and learned that one of his favorite small organic farms was going out of business. Until then, he'd thought they were doing fine, but in reality they were not turning a profit.

Having a successful farm involves more than knowing how to grow crops or raise animals. Farmers also need skills in accounting, marketing, sales, and distribution to create a profitable business. During its first three years, KTA worked with ten farms that, on average, realized a net income increase of 60%. Collectively, the farmers who worked with KTA increased their sales by $1 million each year.

Financial support for KTA's work comes from individuals, foundations, tech companies like Adobe, benefit corporations like Patagonia, and food businesses such as organic groceries, organic produce distributors, and high-end restaurants.[108] This funding allows experts like KTA regional director Thomas Nelson to provide strategic advice and skill building at no cost to farms that are beyond the initial startup phase.

One client Nelson has been working with for three years is Beet Generation Farm in Sebastopol. Libby Batzel and Ali Levesque at Beet Generation started out growing organic produce to sell at farmers' markets, but selling produce directly to consumers was not helping them become financially sustainable. With Nelson's help, they developed a high-value line of vegetable pastas made from their farm's organic produce—beets, tomatoes, spinach, garlic, poblano peppers, serrano peppers, or hot chilis—to increase their revenue. To make the pastas, Batzel and Levesque puree the vegetables, combine them with flour, and extrude them into beautiful pastas with curvy edges that hold sauces well. Nelson explained the benefit of creating value-added products like these, saying, "It's brilliant. They take a perishable product, make it shelf-stable, and then they have cash flow in the off-season."

Since marketing is a key part of business success, Nelson connected

Batzel and Levesque with a food packaging design and brand company called Sloat Design Group. Each year, Sloat chooses one project and provides $50,000 worth of packaging design work *pro bono*. One year, Beet Generation was chosen and received new packaging design that helped them increase their produce sales. Nelson also connected the farm with experts in the sale of package products in stores, helped them set revenue goals, and then figured out how to help the farm reach those goals with an updated business plan. Each year Beet Generation has worked with KTA, they have moved closer to their business goals.

Jobs needed:

- Consultants knowledgeable about business aspects of small organic farms
- Communications experts able to tell stories that close the gap between eaters and the small organic farms who feed them
- Relationship-builders who are able to generate the financial and social capital to fund this work

CHAPTER HIGHLIGHTS

- Forty percent of food grown and raised is thrown away in the U.S.
- Projects to salvage more ugly produce or surplus prepared food could provide jobs to people with entry-level skills.
- Helping small organic farmers with business planning and financial management will help cultivate more small farms that have a circular flow of materials and do not generate waste.

8

MEANINGFUL JOBS

TO RESTORE NATURE

"The sun shines not on us but in us,
the rivers flow not past but through us."
—John Muir

D r. Robin Wall Kimmerer at the State University of New York's College of Environmental Science and Forestry was once leading a graduate writing workshop on people's relationship to the land. During the discussion, students said that nature was the place where they experienced the greatest sense of belonging and well-being. Unreservedly, they claimed they loved the earth.

Citizen Potawatomi Nation member and ethnobotany professor Kimmerer followed up with the question, "Do you think that the earth loves you back?"

No one answered.

Sensing hesitation, she then inquired, "What do you suppose would happen if people believed this crazy notion that the earth loved them back?"

The floodgates opened; they all wanted to talk at once. One person interjected, "You wouldn't want to harm what gives you love." Kimmerer notes that, "We were suddenly off the deep end, heading for world peace and perfect harmony."[109]

In her book *Braiding Sweetgrass*, Kimmerer writes: "Knowing that you love the earth changes you, activates you to defend and protect and celebrate. But when you feel that the earth loves you in return, that feeling transforms the relationship from a one-way street into a sacred bond." Having a sacred bond with another inspires gratitude, a sense of reciprocity, and the will to protect that which is precious to us.

We all know at a profound level that we depend on this one planet with its complex, interconnected natural systems. The earth is therefore "valuable" to us but not in the way Apple and Amazon have a value of one trillion dollars (each). Rather, natural systems are valuable because they support life on the earth.

Stanford University professor Dr. Gretchen Daily has spent her career in biological research ascribing values to ecosystem services. She has also calculated the economic cost of environmental destruction by showing what it would cost if humans needed to recreate specific environmental services. Looking at the numbers she assigns to each ecosystem service makes a strong case for preserving and restoring natural systems. Some of the valuable services nature provides include:

- Purification of air and water
- Mitigation of floods and droughts
- Detoxification and decomposition of wastes
- Generation and renewal of soil and soil fertility
- Pollination of crops and natural vegetation
- Control of the vast majority of potential agricultural pests
- Dispersal of seeds and translocation of nutrients
- Maintenance of biodiversity, from which humanity has derived key elements of its agricultural, medicinal, and industrial enterprise
- Protection from the sun's harmful ultraviolet rays
- Partial stabilization of climate
- Moderation of temperature extremes and the force of winds and waves
- Support of diverse human cultures

- Provision of aesthetic beauty and intellectual stimulation that lift the human spirit[110]

Many of these complex and interdependent systems have been degraded by human activities but at this point are still reparable. We know what needs to be done to restore the healthy forests, healthy waterways, and healthy soil that provide these ecosystem services. Doing so requires appropriating a much-higher level of resources than we are currently allocating for monitoring, research, planning, design, and restoration. Once we do so, however, the legions of people who will be able to do this meaningful work will receive remuneration that is far more than financial. They will literally be saving the earth.

PIVOT #23—CARBON FARMING

After the Marin Carbon Project (MCP) team applied a quarter-inch of carbon-rich compost to selected grazed grasslands sites, scientists from the University of California, Berkeley measured increases in plant growth above and below ground, as well as an increase in the land's ability to absorb water. Treatment sites in Marin County and the Sierra Foothills of California gained an average of one ton of carbon per hectare. Scientists found that just a one-time application of compost boosted deep root growth, which they predicted would result in carbon sequestration that would last for 30 to 100 years.[111]

When comparing application of compost to grasslands, with the alternative of disposing of organic material as waste, there was an even larger carbon benefit. A life cycle study demonstrated that sequestering carbon, as opposed to letting organic material decompose into methane and be released into the atmosphere, "led to large offsets that exceeded emissions, saving upwards of 55 metric tons of CO_2 per acre, per year."[112]

Carbon farming projects are a largely untapped resource for addressing climate change. They fix two big environmental challenges at the same time: topsoil degradation and carbon emissions. Overall, the U.S. loses topsoil at a rate ten times faster than the rate of soil replenishment. Topsoil loss happens when forests are clear-cut for agriculture or grazing, when heavy rain or wind erodes soil from fallow fields, or agricultural chemicals like pesticides and fertilizers kill micro-organisms. According to a 2006 study from Cornell University, U.S. losses of topsoil totaled 1.7 billion tons per year and cost the American economy $37 billion in annual productivity.[113] Rebuilding topsoil counters these destructive processes.

Topsoil forms the pedosphere, the thin skin of living soil on the top three feet of the earth's surface. It is the outer layer of dirt containing minerals, organic matter, water, air, and living microorganisms vital for the planet's land vegetation and essential for growing food.

TABLE 4: CARBON FARMING TECHNIQUES THAT SEQUESTER CARBON

CATEGORY	PRACTICES AND IMPACTS
Afforestation of farmland	Monocultures or mixed species, including tree crops and multipurpose trees
Crop management	Rotations, cover crops, perennial crops, improved varieties
Tillage and residue management	Reduced tillage, crop residue retention
Water management	Rainwater harvesting and other strategies
Biochar application	Application of biochar for fertility and carbon sequestration
Pasture management	Improved pasture species, fodder banks, etc.
Managed grazing	Stocking densities, improved grazing management, fodder production and diversification
Manure application	Application of manure to cropland for fertility; livestock-crop integration
Livestock feeding	Methane-reducing feed and forage
Manure management	Modified bedding, changed feeds, biodigestion, etc.
Agroforestry	Integration of trees with crops and/or livestock
Mixed biomass production	Productive shelterbelts and riparian buffers, biomass crop integration
Perennial protein and biomass coproduction	Concentrated protein as a byproduct of perennial biomass processing

Fortunately, many carbon farming techniques work to rebuild top-soil and sequester carbon. Eric Toensmeier lectures at Yale University and serves as a Senior Researcher at Project Drawdown, specializing in agricultural climate change mitigation. His book *The Carbon Farming Solution* details agroforestry and perennial crop projects with high potential to mitigate greenhouse gas emissions. Table 4 shows thirteen practices and their relative potentials for carbon sequestration.

POTENTIAL GLOBAL MITIGATION IMPACT	EASE OF ADOPTION BY FARMERS	READINESS OF PRACTICE
Medium	Easy	Ready
Medium	Easy	5-10 years
High	Easy	Ready
Medium	Moderate	5-10 years
High	High	Still under development
Low	Low	Ready
Low	Low	Ready
High	High	Ready
Medium	Moderate	5-10 years
High	Moderate to Easy	Ready
Medium	Moderate	Ready
Medium	Easy	Ready
Medium	Easy	Ready

Source: *The Carbon Farming Solution*

Of all these techniques, the following four offer the highest rates of carbon sequestration.

- **Agroforestry woodlot polycultures**—30 tons of carbon/hectare/year
- **Olive tree woody crop monoculture**—26 tons of carbon/hectare/year
- **Moso bamboo woody biomass monoculture**—20–33 tons of carbon/hectare/year
- **Intensive silvopasture (livestock systems) plus timber**—34 tons of carbon/hectare/year

Some carbon farming projects sequester carbon at a lower rate than these four projects but are still worth doing. In fact, California's Healthy Soils Program is funding several different types of agricultural carbon sequestration projects. In 2018, the Healthy Soils Program provided $8.5 million total for grants of up to $75,000 per project for cropland management, compost application, herbaceous establishment on cropland, woody cover on cropland, and grazing lands. Funding for this grant program comes from California's Cap and Trade program, which levies fees on large emitters of greenhouse gases in the state.

Clearly, the work of carbon farming requires human labor, and that costs money. According to the Marin Carbon Project, the cost to sequester carbon through the application of compost to grasslands is about $40/ton. While this is expensive compared to many carbon offset projects, which charge about $10–$15/ton, carbon farming actually sequesters carbon instead of offsetting it. Many carbon offset projects, such as capturing landfill methane, capturing methane emissions from manure with anaerobic digesters, building wind farms, and capturing methane from abandoned coal mines, are all vitally important projects, but they are reducing the growth of global greenhouse gas emissions, not sequestering them.

Our soils can be put to work to help reverse climate change. To do that, we need more carbon sinks that will remove carbon from the atmosphere and bind it up. Jenny Lester Moffitt, who oversees the soil program for the California Department of Food and Agriculture, shared the insight that, "We often look to places like the Brazilian rainforest as carbon sinks, but there's an opportunity to sequester carbon in the soil of our farms and ranchlands."[114]

Jobs needed:

- Planners for the establishment of carbon farms
- Contractors to produce and apply compost
- Carbon farming managers

PIVOT #24—RESTORING HEALTHY FORESTS

The 2017 fire season was one of the worst on record, with damage estimates in the U.S. totaling $18.5 billion. Fires scorched more than 8.15 million acres in the ten states that suffered the most damage: Montana, Nevada, California, Texas, Oregon, Idaho, Alaska, Oklahoma, Kansas, and Arizona.[115] Of this total, 2.3 million acres of national forests managed by the U.S. Forest Service (USFS) burned.[116] As of the writing of this book, the 2018 fire season just ended, with wildfires having burned 30% more land area than the average over the past decade.[117]

In November 2016, the Forest Service conducted aerial flyovers to assess forest health in California. They estimated 102 million trees died during the Golden State's prolonged drought between 2010 and 2016, with about 62 million perishing in 2016 alone. In 2017, the number of dead trees grew from 102 million to 129 million.[118] Persistent drought has made trees even more vulnerable to bark beetle infestations. Western pine beetles and mountain pine beetles have ravaged hundreds of millions of trees in the U.S. and Canada, burrowing into debilitated trees with soft spots and laying their eggs, thus speeding up tree mortality. Raging wildfires then feed on dead or dying trees, the lower canopy of stressed forests, and dry, decaying plant material on the forest floor.

For those wondering what causes wildfires in the U.S., a *Vox* article points to humans. "We fuel them. We build next to them. We ignite them." In California, 645,000 houses are located in high-severity wildfire zones near mountains, forests, and grasslands.[119] Hotter, drier weather in summer and fall, fueled by climate change, turns large parts of the West into tinderboxes sitting vulnerable to an ignition source. Eighty-four percent of wildfires are started by human accidents or activities: downed power lines, careless campfires, vehicle fires, and arson. Once wildfires start, fuel, wind, and long-term dry conditions help them spread quickly.

The burning of 2.3 million acres of national forests in 2017 took a toll on natural areas that the USFS oversees. The USFS is part of the U.S. Department of Agriculture, which manages 193 million acres of national forests for multiple purposes: timber harvesting, recreation, grazing, wildlife conservation areas, fishing, and more. Some of the land managed by the Forest Service serves the purpose of growing timber to be sold on the private market. The Forest Service has a mission encompassing diverse needs to "sustain the health, diversity, and productivity of the Nation's forests and grasslands to meet the needs of present and future generations."[120] Compare this with the National Parks System, a part of the Department of the Interior, which has a lighter touch on the 84 million acres it manages. There are visitor centers, amenities, and roads in national parks, but most of the land the National Parks System oversees remains untouched.[121]

Maintaining and restoring the health of national forests has become curtailed as firefighting increasingly cuts into the USFS's restoration budget, a practice described as "fire borrowing." In 1995, firefighting used 16% of the USFS budget; in 2017, firefighting consumed $2.4 billion, over 50% of the USFS budget.[122]

To help address fire borrowing, Congress passed the fiscal year 2018 Omnibus Spending Package that changed the way the federal government pays for wildfire fighting. Language attached to the 2018 omnibus appropriations bill allows spending cap exemptions to deal with destructive fire seasons, starting in fiscal year 2020, which will prevent "fire borrowing" in the future.[123] This is helpful but does not address the fact that the Forest Service lacks the funding to actively manage the nation's forests at a level that will restore forest health. The impasse owes in part to the different solutions preferred by Congressional Republicans and Democrats.

Republicans would like to open up forests to more logging, with shorter environmental impact analyses for certain fire prevention projects, such as removal of more downed trees near power lines and installation of fuel and fire breaks to stop fire from spreading. Keep in mind that CalFire, Pacific Gas & Electric, and their contractors already remove dead trees that threaten to fall on roads or power lines.

Meanwhile, Democrats are concerned that expanding logging activities would mean rolling back protections provided by the Endangered Species Act and National Environmental Policy Act.

While this impasse continues, timber appraisal specialists like Matt Enzenhouser in Forest Service Region 2 have seen a declining quality of timber harvested from the national forests. Enzenhouser explained that most of the timber coming out of the Rocky Mountain region is salvage timber at this point. Salvage timber, or wood that is gray from bark beetle damage, blue from fungus, or just dry and cracked, fetches lower prices during national forest timber sales.

Timber is sold from national forests in increments of 100 cubic feet (one CCF). To envision one CCF of wood, picture a 100' tree twelve inches in diameter at breast height, where it is cut. Currently, most timber from public lands in Colorado and southern Wyoming sells for $6–11/ CCF. Enzenhouser reports that timber from the Black Hills fetches more than timber from Colorado and southern Wyoming: $25/CCF versus $6–$11/CCF. Forest Service staff would like to have a higher budget for forest restoration work to improve the health of the national forests.[124] For example, some managed forests are so densely packed with trees that the endangered spotted owl cannot move about or spread its wings as it pursues prey.

With additional funding, the Forest Service could hire more people to help make national forests more hospitable to wildlife, fulfilling roles such as:

- **Researchers**—forestry, biology, botany, and archaeology scientists to study the previous and current state of national forests; forest mensurationists; and inventory scientists
- **Planners**—specialists who study the gap between the current state of forests and healthy forests based on input from researchers and then plan how to restore forests to full health
- **Forest technicians**—"boots on the ground" who restore forests by thinning dead trees, removing biomass that could otherwise fuel future wildfires, conducting erosion control, planting seedlings, and removing some roads while building others

Additional resources for researchers and planners are critical in the face of a changing climate. Joe Sherlock, a Regional Silviculturist with the USFS, explained the difficulty of trying to figure out which trees should be planted in the future, given what scientists know and don't know about climate change. Should they be planting trees better suited to drier conditions in the West, and what can they do to ensure these trees survive? Given limited forest restoration budgets, the Forest Service focuses on controlled burns to clear out underbrush, and they would like to be able to selectively cull more dead or dying trees to remove fuel for future potential wildfires but to do so in a way that minimizes environmental impact.

Dr. Camille Stevens-Rumann, a researcher at Colorado State University who specializes in post-fire forest regeneration, explained that a conifer exhibiting orange needles has died within the last twelve months. Forest managers have a one- to two-year window to turn a dead tree into lumber before the wood is too dry to use. Once the moisture content falls below a certain level, the tree splits and cannot provide viable lumber.[125]

Regulatory requirements impede the Forest Service's ability to extract dead or dying trees. The National Environmental Policy Act (NEPA) requires those who want to log in national forests to conduct an environmental impact report before logging. NEPA covers national forests, and the California Environmental Quality Act (CEQA) applies to private forests in California. Both processes seek to protect natural areas by requiring a study about how a proposed activity will affect the natural area in order that decisionmakers can determine if the activity will have a significant negative environmental impact.

However, a catch-22 is at play here: if we have one to two years to turn trees that recently died into wood products for construction or furniture, and both NEPA and CEQA environmental impact reports (EIR) take one to two years to complete, the EIRs run out the clock. By speeding up the EIR process and increasing funding for Forest Service forest restoration work we will be able to reduce fuel loads for the next wildfires.

Mike Albrecht with logging company Sierra Resource Management insists that "we're talking about selective logging, not clearcutting whole forests." According to Albrecht, no clearcutting has taken place in California for over twenty years. When logging happens in national forests, it is done within a few hundred feet of roads and near powerlines. Selective logging is performed to reduce hazards to drivers and potential damage to electric transmission and distribution lines.[126]

To log without clearcutting, logging companies need access to roads and special machinery. Extracting selected trees from the forests involves cutting down a tree, sawing off the branches, and moving it to the road where it is loaded onto a flatbed truck. The speed at which this happens depends on the equipment used. Albrecht explained that one person with a chainsaw can log 20–30 trees in a day; a mechanized system such as a feller buncher can remove 500–600 trees in a day, cutting a wider swath. If you have ever seen video of a feller buncher at work, you will remember well its monstrous, destructive power. This machine clasps onto the base of a tree, cuts the tree at the base, saws off branches with multiple simultaneously spinning blades, then cuts the tree into ten-to-fifteen-foot lengths until all that remains is a stack of logs. One tree can be turned into a pile of logs in less than twenty seconds (do a video search online for footage of a feller buncher in action).

Moving a pile of logs often involves a skidder. This machine, which resembles a construction vehicle, has large rubber tires for navigating forests, usually a few hundred feet at a time. They lift and carry trees out of the forest to the road but generally do not drive miles into the forests. When skidders reach the road with a load of logs, a flatbed truck then delivers them to sawmills.

Sierra Resource Management takes timber to a sawmill like Sierra Pacific industries, the biggest industrial sawmill owner in California, and turns the timber into lumber to be sold.

The Forest Service balances the multiple competing interests of those who seek to leave national forests untouched, the $877-billion-per-year U.S. recreation industry, and the forest and paper products industries. With more resources, the Forest Service could strike the right balance between these interests while restoring healthy forests.

Jobs needed:

- Researchers
- Planners
- Forest technicians

PIVOT #25—CONSTRUCTION PRODUCTS AND FURNITURE

Using some of the severely drought-stressed trees in construction products and furniture, instead of cutting down healthy trees in other forests, would reduce fuel loads for potential wildfires, sequester carbon, reduce pressure to log in healthy forests, and allow more forests to serve as wildlife habitat. Products from higher-value to lower-value use that could be made include:

- **Furniture**—tables, wardrobes, and chairs
- **Construction products**—two-by-fours, structural lumber, oriented strand board, and particleboard
- **Biomass**—scrap wood burned in biomass plants with air pollution controls that remove 98% of the pollution and which generate electricity

Furniture Made from Discolored Wood

Trees visited by mountain pine beetles may suffer gouges or be infected by a fungus that turns the wood blue or another color. While the Forest Service considers this damaged wood to be "salvage" grade, Ohio-based Ghost River Furniture highlights this wood's unique features, such as color and scars, when using it to manufacture tables, chairs, chests, wardrobes, and headboards.

Todd Holbrook, founder of Ghost River Furniture, said, "A lot of woodworkers are afraid to use this lumber. They think they're going to breathe in the fungus and get sick, but it's simply a stain. Another prejudice is that the lumber with any blue stain in it will automatically be Grade 3, the lowest level. But every piece is unique. It's beautiful." Holbrook explained his motivation: "I'm just trying to save the world," then punctuated the sentence with a hearty laugh.[127]

Oriented Strand Board from Salvaged Timber

Oriented strand board (OSB) is an engineered wood similar to particleboard, formed by combining wood strands and adhesives, then compressing the layers in certain orientations. OSB can be made from salvage lumber. Dr. Roy Anderson at forest products planning consulting firm The Beck Group has been studying the economic viability of siting an OSB manufacturing plant in Shasta County in Northern California. While the Western U.S. consumes large amounts of OSB for construction, Figure 26 shows where OSB manufacturing plants currently exist in North America.

FIGURE 26: LOCATIONS OF EXISTING U.S. AND CANADIAN ORIENTED STRAND BOARD MANUFACTURING PLANTS

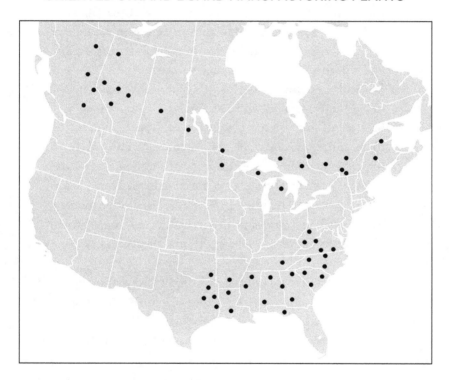

Source: California Assessment of Wood Business Innovation
Opportunities and Markets

Given Shasta County's proximity to forests and the fact that there is no OSB manufacturing plant within 1,000 miles of California, the Beck report studied the business case for making OSB at the former Roseburg Forest Products sawmill site in Anderson, California, for markets in California, Oregon, Nevada, and Arizona.

The study considers three potential sources for raw material that will use parts of a tree other than those used to make structural lumber:

- **Topwood**—A tree's diameter tapers from the butt to the top of the stem. When the diameter of the tree stem falls below six inches, the remainder of the tree becomes too small for use as a sawlog, but it can be utilized as raw material for OSB.

- **Pulpwood**—Pulpwood-size trees are typically no larger than ten to twelve inches in diameter at breast height. These are trees too small to convert into lumber and are typically harvested during thinning, forest restoration, and wildfire hazard reduction treatments.

- **Sawmill byproducts**—Roughly half of the cubic volume of logs entering a sawmill are converted into lumber. The balance of log volume becomes mill byproducts in the form of bark, sawdust, planer shavings, or chips. Of this byproduct volume, roughly half are chips. Chip sales provide a significant source of revenue for sawmills. California, with no remaining paper mills, has limited markets for chips. Therefore, sawmills in California would probably have feedstock to sell to OSB manufacturers.

OSB is made from topwood and pulpwood after removing the bark and branches. In the manufacturing process, knives cut the wood into large "flakes," after which sawmill byproducts and an adhesive binder are added. The mixture is then formed into planks or boards and heated to 500° to dry, thereby creating a solid wood product. The resultant construction material serves multiple purposes in a building—wall sheathing, sub-flooring, and roofing underlayment—which shows how an average residential construction project will use 8,621 square feet of 3/8" OSB.[128]

The Beck report finds that an investment of $166.2 million would be needed to generate 475 million square feet of OSB per year. In this scenario, an estimated $23.57 million of annual cash flow and a 14%(!) annual return on capital would provide a simple payback on the initial investment of 9.6 years. For this project to make financial sense, the OSB manufacturer would need 669,000 green tons of raw material per year.

One issue that needs further development with respect to OSB has to do with the adhesive that binds the chips together in the manufacturing process. OSB predominantly uses synthetic, petroleum-derived thermosetting adhesives, which are mainly based on the reaction of formaldehyde with urea, melamine, phenol, or co-condensates. Formaldehyde is classified as a "known human carcinogen" by the International Agency for Research on Cancer, so finding a less-hazardous binder would increase the chances of an OSB manufacturing plant receiving permitting approval from the state.

Researchers have found bio-based alternatives to formaldehyde such as lignins, tannins, starches and proteins. According to Venla Hammila, who serves as Project Leader at Ikea and who has a Ph.D. in Wood Science and Wood Products from Linnaeus University in Sweden, price and durability pose challenges for the bio-based adhesives currently available. OSB must be durable in moist environments, and biodegradable adhesives are, obviously, biodegradable. At the same time, for OSB to be financially competitive, they must use adhesives that do not cost far more than the conventional option. The bio-based adhesives currently available are not available at the price or quantity the industry requires for mass production.[129]

The main challenge for this project involves commercialization of a non-hazardous binder for OSB at a price-point competitive with the conventional binder. Given California's stringent regulations pertaining to hazardous materials and emissions, the OSB industry could, once it has achieved such standards, potentially create about 160 jobs per OSB manufacturing plant in the Western U.S.

Jobs needed:

- Planners
- Researchers
- Green chemistry researchers
- Forest technicians
- Sawmill and OSB manufacturing workers
- Furniture makers
- Marketing specialists

PIVOT #26—RESTORING HEALTHY WATERWAYS

On April 20, 2010, British Petroleum's (BP) Deepwater Horizon oil rig in the Gulf of Mexico exploded, killing eleven people and creating a fireball that could be seen 35 miles away. Firefighters could not extinguish the fire, and two days after the explosion, the rig sank.

Many of us watched the live underwater camera feed of crude oil spewing into the Gulf day after day as BP and its contractors desperately tried to stop the spill. After several failed attempts, they finally capped the wellhead on July 15, 2010, 85 days after the explosion. To date, this oil spill of 172 million gallons of crude oil remains the largest oil spill in U.S. history.

Emergency response teams deployed 1,740 miles of one-use sorbent booms, 808 miles of containment booms, and 1.84 million gallons of oil dispersants.[130] Operational activity to contain the oil spill and clean up 4,376 miles of shoreline continued for years. At the peak of the cleanup, 48,000 people were involved.[131]

Despite the emergency response and cleanup efforts, the effects of the spill on wildlife were devastating. Immediately after the spill, when state authorities flushed fresh water into the Gulf in hopes of rolling back the oil, oyster beds were wiped out. One year after the spill, at a public meeting in Biloxi, Mississippi, fishermen said they were still hauling up nets full of oil along with their shrimp. In early 2011, more than 150 dolphins, half of them infants, washed ashore, and 87 sea turtles were found dead in one month.

Dr. Samantha Joye, a marine science professor at the University of Georgia, described the view after the BP oil spill in the Mississippi Canyon, which is ten miles from the well. From her research submarine, she saw the ocean floor coated with about four centimeters of dark-brown muck. Thick ropes of slime draped across coral like cobwebs in a haunted house. Few creatures remained alive. Listless crabs were too tired to flee. "Most of the time, when you go at them with a submarine, they just run. They weren't running; they were just sitting there dazed and stupefied."[132]

As of September 2017, BP had paid $63.4 billion to cover clean-up costs and legal fees.[133] Efforts to remove oil from animals, beaches, wetlands, and the sea floor were minimally effective, and many people who rely on the Gulf for their livelihood have suffered financial losses. It's tempting to think how $63.4 billion could be used to transition society away from fossil fuels rather than for oil spill cleanup and legal fees. That dollar amount could go a long way toward establishing bus rapid transit systems, developing public electric vehicle charging infrastructure, setting up electric scooter sharing programs, and creating transportation management association non-profits.

Waterways perform invisible, thankless work for us. Around the world, freshwater bodies, saltwater oceans, the waterways in between, and the adjacent wetlands provide numerous, different ecosystem services. All these areas act as natural filters and sponges, regulate flood peaks, and provide habitat for plants and animals, among other things. Sometimes, we ask too much of these waterways, pushing them too far and then needing to spend a king's ransom to bring them back. Rather than wait to mobilize forces after a major disaster, we should be working to reduce threats to natural systems as well as restore degraded natural systems so they can better bounce back in the face of shocks.

Chesapeake Bay: Half Way There

The Chesapeake Bay's coastline runs some 8,000 miles around Maryland, Washington, D.C., and Virginia. This 200-mile-long waterway provides food, water, cover, and nesting or nursery areas to more than 3,000 migratory and resident wildlife species. Blue crabs, oysters, osprey, cormorants, perch, bass, diamondback terrapins, and river otters are some of its most famous residents.

Every year since 1990, the Maryland Department of Natural Resources and Virginia Institute of Marine Sciences have conducted a Winter Dredge Survey to determine the size of the adult female blue crab population in the Bay. Biologists capture, measure, record, and release crabs from 1,500 sites throughout the Chesapeake Bay to develop their population estimates. Figure 27 shows the steep population declines and rebounds since 1990.

FIGURE 27: CHESAPEAKE BAY ADULT FEMALE
BLUE CRAB POPULATION

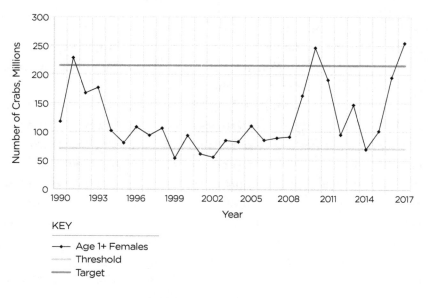

KEY

—●— Age 1+ Females
············ Threshold
——— Target

Source: Maryland Department of Natural Resources

The rebounds testify to adaptive and effective fishery management by Maryland, Virginia, and the Potomac River Fisheries Commission to ensure a stable blue crab population.

Oyster populations have not fared as well. The oyster is a keystone species, which means it is exceptionally important to its ecosystem; in the Bay, oysters purify the water by straining out algae, their food source. Between the 1950s and the 1970s, the annual catch for oysters was 25 million pounds per year. Their population has fallen 98% since then.[134]

The three main threats to the health of the Chesapeake are the nitrogen, phosphorus, and sediment pollution that wash into the Bay. The majority of nitrogen and phosphorus pollution comes from sewage treatment plants, animal feedlots, and polluted runoff from cropland and urban and suburban areas. Agriculture contributes the greatest share of pollution, roughly 40% of the nitrogen and 50% of the phosphorus.

Since the 1970s, as agriculture in Maryland has consolidated into fewer large operations and animals have been confined to smaller spaces, water quality in the Bay has suffered. In the 1970s, the mantra "get big or get out" drove the shift from 4,000 dairy farms with many pasture-fed cows to the current 500. Some farms that used to be dairies now grow corn and soybeans to feed cows. Part of this shift resulted from federal subsidies for commodity crops like corn and soybeans and subsidies for manure management systems, which are needed on CAFOs. In recent years, a massive boom in poultry farming has taken place in Maryland. Soils oversaturated with phosphorus from chicken litter also run off into the Bay.[135]

Nitrogen and phosphorus from fertilizers, cow manure, and chicken litter feed algal blooms in the Bay, which block sunlight to underwater grasses. The bacteria that decomposes algal blooms consumes oxygen in the water, which in turns creates fish-stressing "dead zones." The algal blooms can also spike pH levels, which stress fish and create conditions spurring the growth of parasites.[136] Work to restore the Bay partly involves reducing the Total Maximum Daily Loads of nitrogen, phosphorus, and sediment flowing to the Bay.

To reduce Total Maximum Daily Loads, the U.S. Environmental Protection Agency and non-profits like the Chesapeake Bay Foundation have for decades been facilitating coordination between the seven subnational jurisdictions in the watershed that flows into the Bay. These areas include the six states of New York, Pennsylvania, Maryland, Delaware, Virginia, and West Virginia, as well as the District of Columbia.

Pennsylvania is a key partner in the cleanup work, as the Susquehanna River provides 50% of the water that flows to the Bay.

The Chesapeake Bay Foundation describes the work of cleaning up the Bay as "half way there," as so much has already been accomplished. Cities and towns have upgraded sewage plants with better filters and storm-water systems. With the help of annual incentives of $18 million from the state, Maryland farmers have planted cover crops like clover and legumes, which extract nitrogen from the atmosphere and fix the nitrogen in the soil.

Virginia's Natural Resources Conservation Fund, Maryland's Chesapeake Bay Trust Fund, and Pennsylvania's REAL agricultural tax credit program also provide financial incentives for farmers of crops, poultry, and livestock to implement best practices such as:

- Planting cover crops
- Tilling soil less

- Applying fertilizer 4–6" underground
- Planting pollinator patches
- Practicing precision fertilizer management
- Conducting edge-of-field practices for wetlands, buffer zones, and bioreactors, which filter nitrogen out of runoff with woodchips
- Nutrient management plans and manure pits
- Specific grazing practices

FIGURE 28: WATERSHED FLOW TO THE CHESAPEAKE BAY

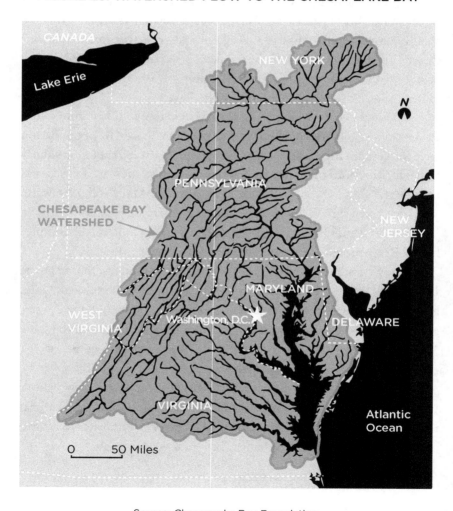

Source: Chesapeake Bay Foundation

Switching from CAFOs to free-range grazing for cows and chickens would reduce the amount of manure running into the Bay. Best practices for grass-fed grazing propose one-half to one acre per cow.

Experts agree that more farmers in the Chesapeake Bay watershed need support managing nutrients in order to further reduce the amount of nitrogen and phosphorus. Technical assistance providers, who go out into the field to ask farmers questions and share information about financial incentives for best practices, provide the help farmers need to make changes that will restore Bay health.

Rob Schnable, a restoration scientist with the Chesapeake Bay Foundation, suggests that diversifying Maryland's agriculture industry to grow more vegetables instead of commodity crops like corn and soybeans would also improve soil and Bay health. He is concerned that the agriculture industry in Maryland could pivot in one of two ways over the next decade. Currently, 25% of all farmers in the U.S. are 70 years old or older. As they retire, their farms might experience further consolidation into large CAFOs or large commodity crop farms and continue being developed into suburban housing developments.

Or, conversely, we could nurture a new generation of eager young farmers who want to grow organic produce and flowers for farmers markets and community-supported agriculture, helping them identify markets for value-added food products from ingredients grown and raised on their farms.

There is also much work to be done in construction site management to reduce sediment running off into the Bay. Best practices include:

- Altering slope and surface roughness via grading and rock dams
- Covering soil with straw rolls, slit fences, or gravel bags
- Planting erosion-preventative vegetation
- Scheduling construction activities during drier seasons and avoiding ground disturbance when water and wind are more likely[137]

Blue crab fishermen have noticed improvements in the health of the Chesapeake Bay in the past few years. As the size of the hypoxic dead zones shrink, the numbers of crabs rebound. Fisherman Scott Wivell's fishing license allows him to operate up to 255 crab pots around Virginia's portion of the Chesapeake Bay. Recently, he only had to check 220 crab pots to reach his daily limit of bushels. "Crabs are everywhere," he said.[138]

Jobs needed:

- Stakeholder coordination project managers
- Outreach technical assistance providers
- Marketing specialists to create demand for organic produce

San Francisco Bay: Tidal Marsh Restoration

Just like in the Chesapeake Bay, pollution and development around the San Francisco Bay have compromised the proper functioning of tidal marshes located near rivers and bays, as well as freshwater inland marshes. Mercury contamination from Gold Rush-era mining, modern salt mining and vehicle pollution, along with flushed medications and wind-blown trash, damage the ability of marshes to absorb water like a natural sponge, accrete sediment, and provide habitat for hundreds of fish and wildlife species.

Before the Gold Rush, tidal marsh area around the Bay totaled 200,000 acres. As of 2007, only 40,000 acres remained. Since then, 5,000 acres have been restored, with another 30,000 acres acquired for future restoration. Another 20,000 acres will bring the total amount of tidal marsh acres up to 100,000,[139] a goal set by the 1999 Baylands Habitat Goals Project, widely embraced by public and private stakeholders and reaffirmed by a 2016 update report.[140]

Several non-profit organizations and government organizations are working together to make this goal a reality. The Environmental Protection Agency pitches in about $5 million per year, and Bay Area voters approved a tax of $12 per parcel, which will provide $500 million for the project over twenty years. The funding is used in part to create jobs, 30 of which are created for every $1 million expended on coastal restoration.[141]

John Bourgeois, Executive Director of the South Bay Salt Pond Restoration project, explained that much of the funding is for restoration work and wishes there was more funding for scientific research. "Potential funders are less enthusiastic about paying for scientific research. This work is really a 50-year experiment, and we're not sure how it's going to work out. There are many aspects of these tidal marshes that we understand very well, but there are also things we don't understand. It would be great to have more funding for the science part of this work."

Still, Bourgeois explained that the funding helps create meaningful jobs that people enjoy. He reported that the people driving earthmoving equipment actually love their work. They know they are helping restore wildlife habitat that will benefit the ecosystem and wildlife that future generations will enjoy.

Jobs needed:

- **Engineers**—to design wetland restoration projects and floor protection levees
- **Scientists**—for design and monitoring
- **Consultants**—for permitting and regulatory approvals
- **Contractors**—to run heavy earthmoving equipment that sculpt sites and construct levees, and nursery and planting crews to grow and install native plants
- **Volunteer coordinators**—to work with thousands of volunteers each year[142]

PIVOT #27—WILDLIFE DEFENSE

Mass extinction of wild animals is one of the most urgent environmental trends currently unfolding. Between 1970 and 2016, Earth lost 60% of vertebrates and 83% of freshwater animals, according to the World Wildlife Fund's (WWF) Living Planet Report.[143] This mass extinction is happening 1,000 times faster than at any time during the past 10,000 years. In the U.S., the extinction of species such as the California grizzly bear (1922), Cascade Mountain wolf (1940), Pallid beach mouse (1959), Smith Island cottontail (1987), and Eastern cougar (2011) are obviously forever.

Mike Barrett, the executive director of science and conservation at the WWF puts the mass extinction into context by noting, "If there was a 60% decline in the human population, that would be equivalent to emptying North America, South America, Africa, Europe, China, and Oceania."

In the WWF's 2018 report, their Global Living Planet Index showed negative trends in population abundance for 16,704 populations representing 4,005 species monitored across the planet between 1970 and 2014. The dotted white line shows the index values and the black parts above and below the dotted white line show 95% confidence intervals.

FIGURE 29: WORLDWIDE SPECIES DECLINE

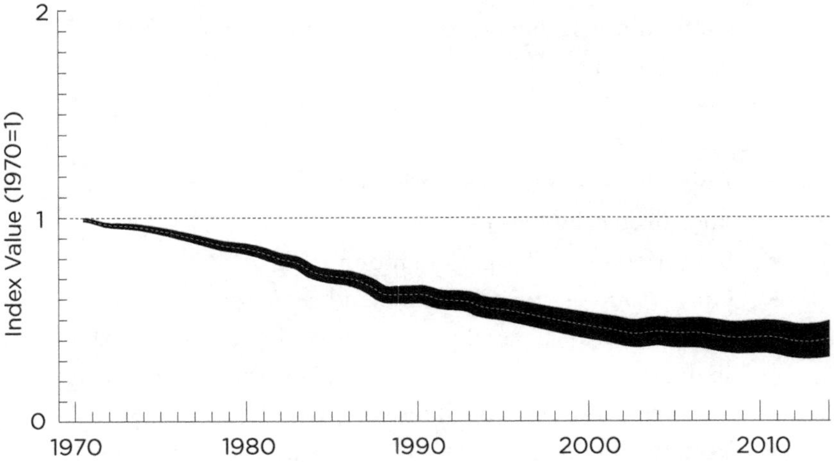

Source: World Wildlife Fund

The previous mass extinction happened 65 million years ago when an asteroid hit Earth. This sixth mass extinction currently upfolding is happening due to a variety of factors. The main threats to wildlife include habitat loss and degradation, species overexploitation, pollution, invasive species, and climate change. Breaking down the threats for different types of species reveals common factors. Habitat loss and degradation pose the greatest challenges for amphibians, reptiles, mammals, and birds.

Barrett underscored the importance of slowing and reversing the trend of species extinction. "This is far more than just being about losing the wonders of nature, desperately sad though that is. This is actually now jeopardizing the future of people. Nature is not a 'nice to have'—it is our life-support system."[144]

Sibley's Guide offers details about factors that contribute to bird mortality in the U.S. Far and away the top-two hazards to birds are

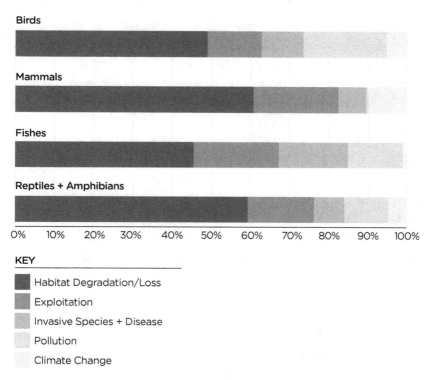

FIGURE 30: CAUSES OF SPECIES DECLINE

North America

Source: World Wildlife Fund

window strikes and domestic or feral cats. The next several hazards that each kill millions of birds, in decreasing order, are high tension power lines, pesticides, cars, communication towers, hunting, and oil and wastewater pits. Lower by orders of magnitude are oil spills, wind turbines, and electrocutions.[145]

Look What the Cat Dragged In

While wind turbines receive negative press for causing bird kills, outdoor cats kill far more birds. According to the Smithsonian Migratory Bird Center, stray and pet cats kill 3.6 million birds every day, on average, in the United States, for a total of at least 1.3 billion birds per year. Is there anything we can do about this slaughter wreaked by our beloved domesticated pets? Sometimes, the answer to a big problem does not need to be complicated.

Birdwatcher Nancy Brennan and bird biologist Susan Willson may have a scalable solution. Living in the Vermont woodlands, Brennan received from her cat, George, gifts of dead bird after dead bird. Tying extra bells to her cat's collar hadn't thwarted his hunting efforts, so when George delivered a struggling ruffled grouse onto the front porch one day, she decided to try something new.

Brennan knew that birds have four color pigments in their eyes, compared to three in primates and just two in other mammals. Using her sewing tools and some multi-patterned fabric, Brennan pieced together what looked like a ruffled Elizabethan collar with a bright color scheme. She fastened it as a cover over her cat's usual collar and let him outside. George returned home later that day with no birds; none over the next few days either. Spring and summer passed without a single dead bird, and she wondered if she was onto something.[146]

Targeted Wildlife Killings

Wildlife is also losing against the protectors of crops and livestock. Targeted killing of animals considered by farmers and ranchers to be a threat results in many wild animal deaths each year. The U.S. Department of Agriculture (USDA) tracks and publishes careful accounts of the number of wild animals USDA staff and contractors disperse, kill, euthanize, remove, or destroy. Below is an excerpt from their 2017 records that includes only those animals that numbered over 1,000 and are not invasive species:[147]

- Beavers—23,644
- Blackbirds, Red-wings—624,845
- Cormorants, Double-Crested—3,155
- Cowbirds, Brown-Headed—285,657
- Coyotes—68,913
- Crows, American—7,346
- Deer, White-Tailed (Wild)—7,524
- Doves, Mourning—22,924

- Ducks, Mallards—2,120
- Egrets, Cattle—4,308
- Foxes, Gray—2,062
- Foxes, Red—1,513
- Geese, Canada—21,488
- Grackles, Common—51,869
- Gulls, California—1,666
- Gulls, Glaucous-Winged—1,906
- Gulls, Herring—3,799
- Gulls, Laughing—5,807
- Gulls, Ring-Billed—4,799
- Hares, Jackrabbits, Black-Tailed—3,954
- Hawks, Red-Tailed—1,701
- Killdeers—3,357
- Larks, Horned—3,012
- Marmots, Yellow-Bellied—1,098
- Meadowlarks, Eastern—1,474
- Meadowlarks, Western—1,198
- Opossums—Virginia—2,832
- Pikeminnows, Northern—65,983
- Prairie Dogs, Black-Tailed—15,233
- Rabbits, Cottontails, Desert—2,798
- Rabbits, Cottontails, Eastern—1,680
- Raccoons—10,313
- Ravens, Common—7,950
- Skunks, Striped—5,291
- Squirrels, Ground, Belding's—4,727
- Squirrels, Ground, California—6,213
- Swallows, Cliff—1,643
- Vultures, Black—7,503
- Vultures, Turkey—1,474
- Woodchucks—1,829

Some people are working to protect some of these targeted animals. Dave Crawford helped form the Prairie Dog Coalition in the early 2000s to bring populations of prairie dogs threatened by development back from the brink of extinction. Prairie dogs are important to specific ecosystems as a keystone species that support an array of other wildlife.

More recently, Crawford founded the Animal Help Now service, which with its app and web portal seeks to be "the 911 of wildlife emergencies," whether helping a bird who has struck a window, an injured deer on the highway, or a prairie dog colony at risk of bulldozing. Animal Help Now empowers citizens by providing resources, strategies, and community. Crawford explained, "Humane removal services are available, but not to the extent needed. More humane wildlife control operators are needed worldwide, and the public needs an easy way to find them."

Other non-profits focus on innovative solutions for saving wild animals at the top of the food chain, specifically bears, wolves, and mountain lions. These groups understand how important apex predators are to the health of an ecosystem. In the Rocky Mountain region, the non-profit People and Carnivores works with ranchers, hunters and outfitters, rural residents, land managers, and scientists to reduce the number of human-predator conflicts. They engage ranchers and farmers by first listening to their concerns, then gauge their level of acceptable risk and talk through potential ideas to prevent conflicts with predators in the area. Some of the tools and best practices used to reduce conflicts include:

- Range riders and livestock management
- Electric fencing
- Fladry fencing (a line of rope mounted along the top of a fence with colored flags that flap in the breeze)
- Livestock carcass composting and management
- Livestock guardian dogs
- Scare devices

These best practices have proven effective in keeping grizzlies, wolves, and other carnivores out of trouble. With more funding, non-profits like People and Carnivores or Predator Defense could do more advocacy work on behalf of wild animals and collaborate with farmers and ranchers to find non-lethal predatory control methods that preserve wildlife.

Jobs needed:

- Outreach technical assistance providers
- Stakeholder coordination project managers
- Grant writers

PIVOT #28—WILDLIFE RESTORATION

Many faiths express the importance of being good stewards of the earth. In the Jewish tradition, the term *tikkun olam*, "repair the world," comes from the rabbinical teaching that a holy vessel containing divine light was shattered when the Holy One created the world. Repairing the world in the modern era means restoring the world to its original state of wholeness through actions that better both one's own community and the well-being of future generations. Those working on sustainability projects that restore climate stability, build resilience into our infrastructure, advance social justice and inclusion, or help wildlife populations rebound are just a few examples of *tikkun olam*.[148]

There is amazing work being done to bring back threatened or endangered species in our national parks. Because they are some of the most protected natural areas in the country, these parks provide critical habitat for conservation work.

The non-profit National Parks Conservation Association touts nine success stories where multiple government agencies and non-profits worked together to help the numbers of a variety of endangered and threatened animals rebound to the point where they could be reintroduced to their native habitats.

- **Pacific fishers (Mount Rainier Park, North Cascades Park, and Olympic National Park, Washington)**— Trapping and logging decimated the populations of Pacific Fishers, which built dens in the old growth forests of the Cascade Mountains. Interventions by the National Park Service, Washington Department of Fish and Wildlife, and the U.S. Geological Survey worked together to successfully rebuild their numbers.
- **Black-footed ferrets (Badlands Park and Wind Cave National Park, South Dakota)**—In 1987, there were only eighteen known black-footed ferrets in existence. They were put in a captive breeding program and reintroduced to Badlands National Park in 1994 and Wind Cave National Park in 2007. About 1,000 of these animals

now live in the wild, thanks to decades of work involving multiple government agencies.

- **Bald eagles (Channel Islands National Park,** California)—Exposure to synthetic pesticide DDT caused the number of bald eagles in the Channel Islands to crash in the 1950s. Between 2002 and 2006, biologists released 61 eagles from breeding populations in Alaska and other parts into the Channel Islands. Today, 40 live there.

- **Gray wolves (Yellowstone National Park, Idaho, Montana, and Wyoming)**—In the mid-1920s, gray wolves were eradicated from Yellowstone National Park and surrounding areas in order to protect cattle populations. Today, we better understand the critical role wolves play in ecosystems. The U.S. Fish and Wildlife Service and National Park Service worked together to reintroduce gray wolves to Yellowstone Park in the 1990s, and today tourists strive to see some of the 100 wolves currently in the park.

- **Desert pupfish (Organ Pipe Cactus National Monument, Arizona)**—Groundwater pumping, which dried up habitat, and non-native species competing with this minnow-like creature led to a decline in its numbers. Park staff have been working to reclaim habitat and reintroduce the species, and today there are about 4–5,000 desert pupfish in the park's Quitobaquito Springs.

- **Bighorn sheep (John Day Fossil Beds National Monument, Oregon)**—Overhunting and diseases caused the demise of bighorn sheep in the park in 1905. In 2010, the National Park Service, Oregon Department of Fish and Wildlife, the Bureau of Land Management, and ranchers released 20 bighorn sheep into the park and nearby lands, where they have since repopulated.

- **California condors (Pinnacles National Park, California)**—In 1982, fewer than 22 California condors remained, due to various human activities and consumption of lead ammunition. Today, after interventions by the U.S. Fish and Wildlife Service, 210 birds live in the wild and 180 in captivity.

- **Elk (Great Smoky Mountains National Park, North Carolina and Tennessee)**—Hunting, trapping, and habitat loss resulted in such low elk population numbers that conservationists were concerned they would become extinct. In 2001, the Park Service reintroduced elk to the Great Smoky Mountains National Park, where 140 now live.
- **Nēnē (Hawaii Volcanoes National Park, Hawaii)**— This native goose population suffered from hunting, habitat loss, and predation by non-native species such as mongooses, cats and dogs. By the 1940s, there were only 50 left. Captive breeding and reintroduction have helped restore healthy numbers.[149]

Further expanding efforts like these by biologists and technicians to restore the numbers of endangered and threatened wild animals will help wildlife be able to self-regenerate. This is a gift to wildlife and to future generations of humans who will enjoy having a biodiverse planet.

Jobs needed:

- Wildlife biologists
- Project managers
- Technicians

PIVOT #29—WILDLIFE OVERPASSES AND UNDERPASSES

Wildlife needs three things to survive: food, shelter, and access to a second breeding population different from its own. Wild animals wandering in search of these things often cross roads, which can be lethal. To address habitat fragmentation from development, more and more states have started building wildlife overpasses and underpasses.

Different types of wild animals have their preference for overpasses or underpasses. Grizzly bears, wolves, moose, and deer usually chose overpasses to cross roadways. Cougars, however, feel more comfortable in underpasses. Some overpass designers have added rock piles to overpasses as pikas and salamanders find them helpful.

The state of Colorado installed five underpasses and two overpasses across Highway 9, south of Kremmling, which reduced wildlife-related crashes by almost 90%. The Washington State Department of Transportation is working on 27 animal crossings, mostly underpasses, for a fifteen-mile stretch of I-90 east of Keechelus Lake. Some features include fences to funnel animals toward the crossing as well as soil, trees, and native plants on the overpasses. High walls on both sides of the crossing reduce the glare of headlights, which can confuse animals. Nevada, Utah, and Wyoming, to name just a few, are also planning wildlife crossings.

Jobs needed:

- Planners
- Designers
- Construction project managers
- Construction workers
- Monitors and maintenance workers

PIVOT #30—REWILDING FOR HABITAT RESTORATION

Biologist Dr. E. O. Wilson proposes setting aside half of Earth for nature. The idea is that by conserving half the land and sea, 85% of all species will be protected from extinction, and life on Earth will enter the safe zone. The project is more theoretical than action-oriented at this point, but the Half-Earth Project has spawned some interesting discussions. Which areas of land should we set aside for nature, and how can we make sure doing so does not come at the expense of those struggling to make ends meet?

To help drive understanding of conservation and inform this policy discussion, the Yale Map of Life project has been mapping species by location around the world. The maps and data sets found on MOL.org show species ranges, inventory, and occurrence. In the Map of Life, one rectangle at the Equator measures 150 kilometers by 150 kilometers. Farther north or south, the rectangles become taller. This scale of the space describes the richness of the species.[150] Each rectangle lists the number of different vertebrate, invertebrate, and plant species. For example, one rectangle in Kentucky contains 276 birds, 40 mammals, 49 reptiles, 15 turtles, 57 amphibians, 209 fishes, 146 butterflies, fifteen bumblebees, 166 dragonflies, 159 trees, and one cactus.

How close are we to the Half-Earth Project's goal of setting aside half of the world's land and half of the oceans for conservation? Globally, humans have reserved just under 15% of the world's land, including inland waters, and just over 10% of the coastal and marine areas within national jurisdiction. In the oceans, 4% of global marine areas are designated Protected Areas.[151]

A study from scientists at Yale University and the University of Grenoble finds that a 5% expansion of protected land could triple the protected range of key species and safeguard their functional diversity. But setting aside land is not as simple as rounding up wildlife and putting a fence around it.

"Some species offer more critical or unique functions or evolutionary heritage than others—something current conservation planning does not readily address," explained Walter Jetz, Director of the Yale Center for Biodiversity and Global Change.[152] Protecting the number of species is important, but so is the phylogenetic diversity, meaning the degree to which the tree of life is well represented. Ensuring functional diversity in which a variety of diet types or body sizes are represented in an ecosystem should also be a goal of conservation planning.

To help wildlife make a comeback, Dr. Wilson would like to see more research done to map critical habitat and species richness, saying, "By determining which blocks of land and sea we can string together for maximum effect, we have the opportunity to support the most biodiverse places in the world as well as the people who call these paradises home."

Marine Wilderness

Only about 13% of the world's ocean, about 55 million square kilometers, are classified as marine wilderness, which describes areas left relatively untouched by humans. Most marine wilderness is located in the high seas of the southern hemisphere and at extreme latitudes where sea ice prevents human access to the ocean. Little marine wilderness remains in coastal areas. Just under 5% of marine wilderness currently lies within marine protected areas.

An article in the research journal *Current Biology* shows a map of the remaining marine and terrestrial wilderness that provide relatively untouched habitat for wildlife.

FIGURE 31: WILDERNESS AREAS

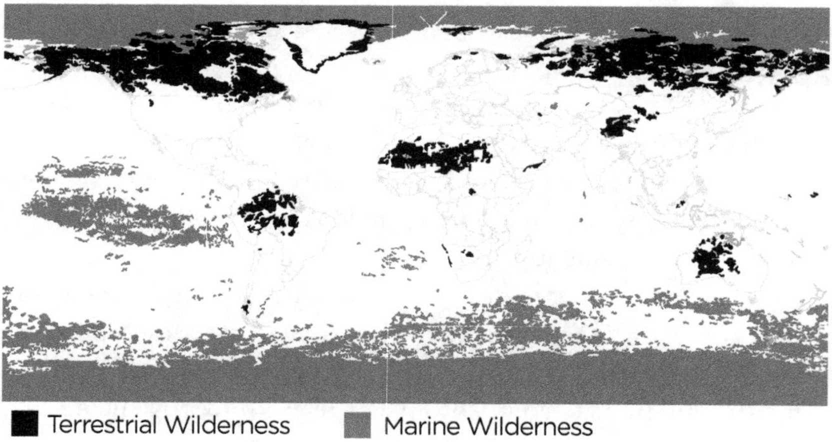

■ Terrestrial Wilderness   ■ Marine Wilderness

Source: *Nature, International Journal of Science*

Wilderness areas contain high genetic diversity, unique functional traits, and endemic species. They maintain high levels of ecological and evolutionary connectivity, and may be well placed to resist and recover from the impacts of climate change.[153]

Multilateral environmental agreements will be key to managing human activities that threaten wild species: overfishing, destructive fishing practices, ocean-based mining that extensively alters habitats, oil drilling, military exercises, cargo shipping, and limiting runoff from land-based activities. Technologies of extractive human activities have extended their reach and impact on wildlife over the past several decades. Fishing gear improvements have increased the mean depth of industrial fishing by 350 meters since 1950. Oil and gas wells operating deeper than 400 meters number around 2,000. Putting agreements in place to designate areas off limits to human activities would allow marine wildlife populations in these areas to rebound.

Wild animals need wild spaces to find the food, shelter, and second breeding population they require to survive. If the sprawl of humanity across previously wild spaces were to be re-concentrated to allow more space for flora and fauna, or if we could find ways to coexist with other species, we could slow down the mass species extinction currently unfolding.

To figure out where to start, let's take a look at how land is used in the U.S. A map from a July 2018 article in *Bloomberg*[154] takes the aggregated data for land use and depicts it on a map to show relative impacts.

FIGURE 32: LAND USE IN THE U.S.

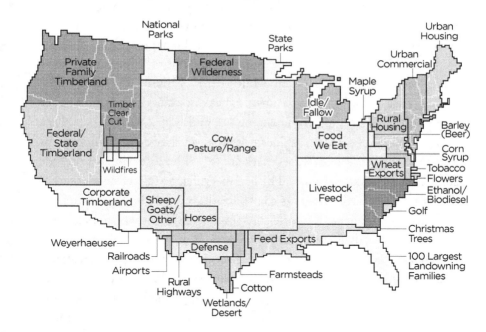

Source: *Bloomberg*

Note that Figure 32 shows that if we combined all federal wilderness in the continental U.S., it would occupy an area smaller than North Dakota and South Dakota put together.

Wilderness was defined by Congress in the 1964 Wilderness Act as "an area of undeveloped Federal land retaining its primeval character and influence without permanent improvements or human habitation, which is protected and managed so as to preserve its natural conditions." With the exception of Alaska, wilderness areas do not allow motorized equipment, motor vehicles, mechanical transport, permanent structures, or installations.

More Public Conservation Land Is Vital to Restoring Endangered Species

Expanding the amount of wilderness that serves as wildlife habitat would give threatened and endangered wildlife more places to regenerate their numbers. Setting aside more land for wilderness would also serve restorative and spiritual benefits to humans. Each year, close to 300 million people visit national parks, mostly in the summer months, and the vast majority of those visitors stay in the parts of the parks with amenities.

The National Park Service explains the value of wilderness areas to humans, even if they rarely enter into them, as thus: "Wilderness provides a sense of wildness. Just knowing that wilderness exists can produce a sense of curiosity, inspiration, renewal, and hope. Most park visitors will probably never enter a wilderness area, yet they enjoy wilderness as a scenic backdrop to developed park areas.[155] These sentiments provide some of the reasons writer and historian Wallace Stegner once referred to our national park system as the "best idea we ever had."

To humans, wilderness offers inspiration and spiritual renewal, but to threatened or endangered wildlife, wilderness areas are vital, safe places. When national park restoration biologists are ready to release threatened and endangered wildlife species back into the wild, they do so within national park wilderness. Setting aside more public land for rewilding will provide space for threatened and endangered species to regenerate.

The stewards of federal wilderness include the National Park Service, U.S. Fish and Wildlife Service, U.S. Forest Service, and the Bureau of Land Management; however, federal budget cuts have hampered the ability of researchers, planners, rangers, and technicians in these agencies to protect and conserve species. To make progress toward the goals of the Half-Earth Project, these agencies need more funding to expand the number of people doing this important work.

Conservation Work on Private Land

In 2017, populations of monarch butterflies overwintering in Mexico fell 14.8% in one year due to severe hurricanes during the monarch migration and unseasonably warm fall weather.[156] Then in 2018, monarch populations in California fell 86% from the previous year. The Xerces Society for Invertebrate Conservation reported that the 2018 Thanksgiving and 2019 New Year's monarch butterfly tallies at 213 sites in California only counted 28,429 orange-and-black butterflies. That's a 99.4% decline from populations in 1980 when an estimated 10 million monarchs migrated from Marin County, California, down to Baja California.[157]

Recent population declines have been so precipitous that the U.S. Fish and Wildlife Service is considering listing them as endangered. The agency will wait to make a determination though, to see how effectively collaborative efforts by farmers, universities, agribusiness companies, and state agencies boost monarch butterfly populations. In Iowa, farm organizations, Iowa state agencies, and Iowa State University are working together to plant habitat for butterflies on less-productive farmland and pastures, in buffer strips and conservation areas, on roadsides, and in community gardens and demonstration projects. If there is no progress by mid-2019, the plan is to officially list monarch butterflies as endangered.

Urbanization as Part of the Solution

Reducing the amount of land on which humanity has an impact will free up more area for wildlife. Part of this encroachment issue has to do with the density of human settlements. According to the 2010 U.S. Census, 80.7% of people lived in urban areas. As more of the younger generations leave rural areas and head for jobs in the cities—an urbanization trend that has continued unabated since the founding of the U.S.—articles decry the depopulation of small towns. But what if urbanization were not a problem but rather part of the solution to the problem of mass species extinction?

Kim Stanley Robinson posited an interesting idea in a March 2018 *Guardian* article titled: "Empty half the Earth of its humans. It's the only way to save the planet." Robinson described a new role rural areas could serve, saying, "If these places were redefined (and repriced) as becoming usefully empty, there would be caretaker work for some, gamekeeper work for others, and the rest could go to the cities and get into the main swing of things."[158]

Jobs needed:

- Biologists to survey and monitor wildlife populations
- Researchers
- Technicians
- Outreach and education specialists
- Stakeholder coordination project managers
- Policymakers
- Planners

Conclusion

In order to graduate, students at SUNY's School of Environmental Science and Forestry must spend five weeks living off the land in the wilderness near Cranberry Lake Biological Station in upstate New York. Professor Kimmerer starts by leading her fifteen or so students each year on a "shopping" expedition of sorts in nature. They gather natural materials to make shelter, bedding, insulation, rope, tools, light sources, food, heat, rain gear, shoes, and medicine.

The first project involves building a wigwam where the class will sleep. Students harvest saplings and set them deep in the soil in a twelve-foot circle for the wigwam's frame. Students bend the tops of the maple saplings and tie them across the circle in arches. When they are done, the frame looks like an upside-down bird's nest. Next, they harvest cattails from a nearby marsh and proceed to weave cattails into shelter walls and cushiony sleeping mats. The cattails also provide leaves to make string and twine and roots for food, and when the part of the stalk with matted fuzz is dipped in fat, it can be lit to make torches. The class unearths spruce roots with which to weave baskets, and they tie birch bark sheets together to make the shelter's roof.

Over the course of the five weeks, the students develop a genuine appreciation for nature's gifts. As their tour guide through the biological wonders to be found in root systems, marshes, and among flowers, Kimmerer also teaches them how to enter into an agreement with the earth that honors the gifts nature gives us. Students come to understand the basic tenets of the agreement over the course of the field trip, which are:

- Ask permission before taking. Abide by the answer.
- Never take the first. Never take the last.
- Take only what you need.
- Take only that which is given.
- Never take more than half. Leave some for others.
- Harvest in a way that minimizes harm.
- Use the harvest respectfully. Never waste what you have taken.
- Share.
- Give thanks for what you have been given.[159]

As we remember to feel gratitude for the gifts we harvest, the reciprocity that flows from this gratitude encourages us to find ways to heal and restore the earth.

CHAPTER HIGHLIGHTS

- Proofs of concept exist for restoring topsoil, forests, waterways, and wildlife that could be replicated and scaled.
- Carbon farming provides opportunities to sequester carbon while giving farmers and ranchers additional income streams.
- Finding ways to shrink humanity's footprint and coexist with wildlife will give wildlife the space it needs to survive and thrive.

PART II: THE PLAN

9

DISRUPTING BUSINESS AS USUAL

"Be bold and mighty forces will come to your aide."
—Goethe

B ack in 1972, Club of Rome researchers hypothesized that humanity was on an unsustainable trajectory. Dennis Meadows, Donella Meadows and their colleagues developed a computer model, World3, which predicted how global environmental trends would play out over the next several decades. Their research sought to better understand what would happen to Earth and its inhabitants if the then-current growth rates of the food system, industrial system, population, non-renewable resource use, and pollution continued. Researchers tweaked the model over a few iterations, and, in 1972, the Club of Rome published the book *The Limits to Growth*, which described the World3 model's results. World3 predicted the collapse of civilization around 2072 if global environmental trends continued.

Criticism of *The Limits to Growth* was swift, strident, and widespread. After the fervor died down, the world settled back into business as usual: accelerating population growth, steadily growing per capita resource consumption, and continued generation of pollution.

Then, in 2014, Graham Turner, a research fellow at the Melbourne Sustainable Society Institute, revisited the World3 model. He and his colleagues gathered updated data on population, natural resource consumption, pollution, and other factors from the United Nations, the National Oceanic and Atmospheric Administration, and other agencies. Turner laid the actual statistical data next to the decades-old projections and found the new data tracked quite closely with World3 model through 2010.

While Turner notes that his team's research validated World3 data, he and his colleagues offered a small window of hope in their findings.

"Our research does not indicate that collapse of the world economy, environment and population is a certainty. Nor do we claim the future will unfold exactly as the MIT researchers predicted back in 1972. Wars could break out; so could genuine global environmental leadership. Either could dramatically affect the trajectory. But our findings should sound an alarm bell. It seems unlikely that the quest for ever-increasing growth can continue unchecked to 2100 without causing serious negative effects—and those effects might come sooner than we think."[160]

Business as usual is leading civilization straight toward a cliff, and disrupting this trajectory would save us from hurtling off its edge. In the tech business, disruption is generally considered to be a good thing. Startups—whether in Silicon Valley, Silicon Prairie, or Silicon Alley—look for a business niche to disrupt. Wherever established companies are providing goods or services that are expensive, inefficient, and not meeting the needs of society, startups identify opportunities to deliver better, faster, and cheaper alternatives. Using a similar approach, we need small, nimble organizations to disrupt companies in extractive, non-renewable, and wasteful business sectors. At the same time, we need to support individuals within these dinosaur business sectors to evolve into more sustainable enterprises.

There are people within these organizations working top-down and bottom-up to become more sustainable. CEOs and Board of Directors at publicly-traded companies would like to be able to consider more than just the next 90 days of profits in their decision-making. Many have said they would like to do more long-term planning, but they are constrained by calls to maximize stock price and shareholder value.

Then there are often people who would like to green their organization's operations or offer greener products and services. In 2008, the Sustainability Manager for Clorox spoke at the Sustainable Brands conference to highlight the company's then-new GreenWorks line of bio-based household cleaning products. During the question-and-answer section of the talk, an audience member asked why Clorox didn't green their entire product line, thereby making all their products less hazardous and better for the environment. The Sustainability Manager let out a little sigh and explained that Clorox is a publicly traded Fortune 500 company. Their new GreenWorks line of household cleaners was a trial balloon: they wanted to see if people were compelled to purchase them. If the products were popular, the company would slowly expand their offerings. This was, he said, an incremental process. They couldn't change all of their products overnight, because if investors did not like the swift pace of change, the company's stock price might suffer a precipitous drop.

Many visionary change agents, like Clorox's Sustainability Manager, are nudging their conventional companies in a more sustainable direction. They often face strong headwinds, but they play a key role in helping the business community evolve. One vital way we can support them is by having a conversation as a society about the goals of our economy. Can most of us agree that everyone's basic needs should be met, that everyone should have food, clean water, sanitation, shelter, education, and access to healthcare? Can we also agree that we should rein in the ways we are overshooting planetary boundaries: climate change, biodiversity loss, nitrogen and phosphorus loading, and land conversion? Again, our goal must be to shift humanity to the safe space between the social foundation and the ecological ceiling described in *Doughnut Economics*.

As we become clearer about the goals of our economy and our society, we also want to be mindful of the metrics we use, as the current primary metrics do not help us move into that safe space. The metrics we hear about regularly in business news are stock market indicators, the gross domestic product (GDP), and the unemployment rate. These metrics, which were once helpful in measuring certain types of improvement, are now anachronistic and not useful in reversing climate change and species extinction. Worse, in some cases, they leave out such vital elements as to be misleading.

QUESTIONING EXISTING METRICS

Kate Raworth in her book *Doughnut Economics* explained the history of the development and use of the GDP in order to demonstrate why it's no longer a useful indicator of progress. In the mid-1930s, Simon Kuznets devised GDP as a measure of America's national income. The metric assigned a dollar value to America's annual output so decisionmakers in Washington, D.C., could compare it to the year before. President Franklin Delano Roosevelt used it to monitor the changing state of the U.S. economy and assess the impact of his New Deal policies to pull America out of the Great Depression.

As the country entered World War II and the U.S. converted its competitive industrial economy into a planned military economy, data underlying the GDP accounts helped policymakers track whether production was meeting society's needs. Then, amidst ideological competition between the Soviet Union and the United States in the 1950s, the GDP proved useful in showing whose economic ideology could create more stuff: the free market or central planning.

For decades, economic growth was seen as a way to end unemployment. Arthur Okun, Chairman of President Lyndon B. Johnson's Council of Economic Advisers, developed a correlation between national output and unemployment, which became known as Okun's Law. The law found that a 2% growth in U.S. national output corresponded to a 1% fall in unemployment. "Growth was soon portrayed as a panacea for many social, economic, and political ailments: as a cure for public debt and trade imbalances, a key to national security, a means to defuse class struggle, and a route to tackling poverty without facing the politically charged issue of redistribution."[161]

During John F. Kennedy's presidential campaign, Kennedy promised a 5% economic growth rate. After he won the election, he turned to his chief economic adviser and asked, "Do you think we can make good on the five percent growth promise?"[162]

For decades, GDP has been the primary measure of progress, even though it has blind spots. At its essence, GDP is a measure of money changing hands. Some transactions that count toward the GDP do not make society better off. Every vehicle accident that puts someone in the hospital and every oil spill that necessitates cleanup increases the GDP, but this does not mean society derives benefits from the car accident or oil spill. The economist who invented the GDP as we know it, Simon Kuznets, understood the indicator's shortcomings and warned us not to confuse GDP with "economic welfare." Robert F. Kennedy liked to say during his 1968 presidential campaign that GDP measured everything "except that which makes life worthwhile."

Another flawed economic metric we commonly use is the Bureau of Labor Statistics' U-3 unemployment rate. As mentioned in the opening chapter, the major blind spot of the current unemployment rate of 3.7% is the fact that it ignores the 37 million people between the ages of 25 and 64 who are Not in the Labor Force; it only measures the people who actively looked for work in the past four weeks. By using the U-3 unemployment rate, we ignore the tens of millions of people who would like to be contributing members of society and be offered meaningful work but are locked out of the work world for various reasons. Using such a misleading metric hinders job creation. Since we are told that there are more open jobs than people looking for work, we do not feel any urgency to create more work.

APPLYING NEW METRICS

We need to redefine success with different metrics. We should be measuring success by how happy people are. The World Happiness Report[163] could be our new standard. The Global Happiness Index in the World Happiness Report measures six variables:

1. Gross domestic product per capita
2. Social support
3. Healthy life expectancy at birth
4. Freedom to make life choices
5. Generosity
6. Perception of corruption in government and in business

Regarding the second variable, to determine a person's degree of social support, researchers asked, "If you were in trouble, do you have relatives or friends you can count on to help you whenever you need them, or not?" To measure the fifth variable, generosity, the report asks if "you have donated money to a charity in the past month." Overall, these six variables allow us to compare relative happiness in various countries around the world. Researchers interviewed people in 156 countries and found the self-reported happiest people rank in the following descending order.

The five Nordic countries and two other European countries occupy seven of the top ten spots in the world's happiest countries. Note that these countries have strong safety nets and that only one of the six variables in the World Happiness Index is income.

Drilling down on just the income-per-capita metric, another study asked 1.7 million people from 164 countries about the optimal salary they would need to achieve fulfillment. The average response around the world was $95,000. In North America, survey respondents shared that the optimum salary was $105,000. Researchers found that beyond this point, the benefits of making more money decreases. Keep in mind that the $105,000 number is for an individual, while the average household income for the U.S. is $65,000 and 75% of American households earn less than $75,000.

Andrew Tebb, the study's lead author, found that "increases in happiness tend to diminish as you make more money." Ultimately though, "Income is just one variable in the complicated equation of happiness. It's not trivial, but there are other factors that are at least as important, such as meaning and significance and social relationships, family, and friends," Tebb said.[164]

We cannot continue economic growth as long as economic growth means consumption. We are used to an annual economic growth rate of 3%, which means doubling the size of the economy every twenty years. If the fact that our economy is growing reflects how much more we are consuming, then every twenty years, we will double the number of cars we buy, the amount of fish we catch, the amount of minerals we extract, and the number of iPhones we purchase—then double it again in another twenty years. For us and this planet to survive, we need to decouple economic growth from material throughput.

FIGURE 33: COUNTRIES WITH SELF-REPORTED HAPPIEST PEOPLE

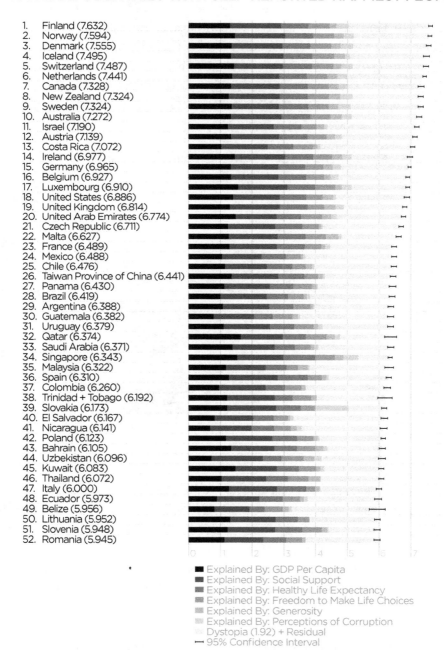

1. Finland (7.632)
2. Norway (7.594)
3. Denmark (7.555)
4. Iceland (7.495)
5. Switzerland (7.487)
6. Netherlands (7.441)
7. Canada (7.328)
8. New Zealand (7.324)
9. Sweden (7.324)
10. Australia (7.272)
11. Israel (7.190)
12. Austria (7.139)
13. Costa Rica (7.072)
14. Ireland (6.977)
15. Germany (6.965)
16. Belgium (6.927)
17. Luxembourg (6.910)
18. United States (6.886)
19. United Kingdom (6.814)
20. United Arab Emirates (6.774)
21. Czech Republic (6.711)
22. Malta (6.627)
23. France (6.489)
24. Mexico (6.488)
25. Chile (6.476)
26. Taiwan Province of China (6.441)
27. Panama (6.430)
28. Brazil (6.419)
29. Argentina (6.388)
30. Guatemala (6.382)
31. Uruguay (6.379)
32. Qatar (6.374)
33. Saudi Arabia (6.371)
34. Singapore (6.343)
35. Malaysia (6.322)
36. Spain (6.310)
37. Colombia (6.260)
38. Trinidad + Tobago (6.192)
39. Slovakia (6.173)
40. El Salvador (6.167)
41. Nicaragua (6.141)
42. Poland (6.123)
43. Bahrain (6.105)
44. Uzbekistan (6.096)
45. Kuwait (6.083)
46. Thailand (6.072)
47. Italy (6.000)
48. Ecuador (5.973)
49. Belize (5.956)
50. Lithuania (5.952)
51. Slovenia (5.948)
52. Romania (5.945)

Explained By: GDP Per Capita
Explained By: Social Support
Explained By: Healthy Life Expectancy
Explained By: Freedom to Make Life Choices
Explained By: Generosity
Explained By: Perceptions of Corruption
Dystopia (1.92) + Residual
⊢—⊣ 95% Confidence Interval

Source: World Happiness Report 2018

Redefining success as happiness is one way to accomplish this. The foundation of such a redefined metric includes: making sure our basic needs are met, knowing there is a safety net available to catch us if we fall, knowing we have friends and family to support us, and, absent that, a bare minimum of financial support to sustain us until we can get back on track.

Raworth explains the pathways of resource use that would allow us to decouple growth from GDP and move below the limits of planetary boundaries, as shown in Figure 34.

FIGURE 34: DECOUPLING GROWTH FROM GDP

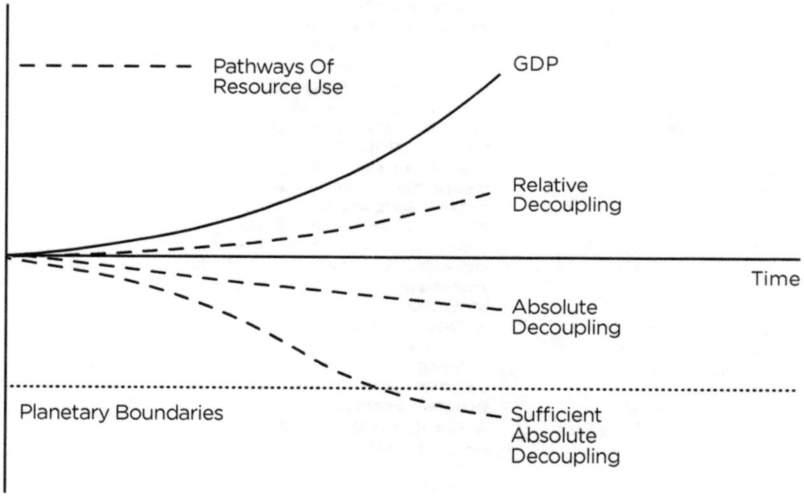

Source: Kate Raworth, *Doughnut Economics*

To achieve sufficient absolute decoupling, our society must be willing to do four things:

- Shift from fossil fuels to renewables
- Create a circular flow of materials
- Dematerialize by shifting to digital products and services
- Radically reduce waste

The last bullet point seems like a no-brainer. Why wouldn't we reduce waste that provides no value? Abundant opportunities exist in the reduction of energy, water, materials, and food wastage. Retrofitting for energy efficiency and water conservation and reducing food waste and diverting surplus prepared food to those who are food insecure do not involve sacrifice.

Instead of focusing on growing the GDP, we should change our goals to make sure that:

- Everyone's basic needs are met
- We restore natural systems to the point that they can regenerate themselves
- We leave a livable planet for future generations

These points form a new social contract to guide us. One company has already incorporated the last goal, to leave a livable planet for future generations, into their board's decision-making processes. When their board of directors meets, they leave one chair empty to represent future generations. Whenever the board votes on an issue, the board members contemplate what future generations would think about the issue and then give them one vote.

To accomplish the goals of a new social contract, the private, non-profit, and government sectors all have vital roles to play, as each sector embodies strengths that make them well-positioned to help civilization move into the safe space between the social foundation and planetary limits. The private sector can quickly deploy abundant resources to complete well-defined tasks. The non-profit sector is mission-driven and operates in a lean manner. Government serves the public good and can be the vehicle through which the will of the people becomes manifest.

In fact, our government is uniquely positioned to accomplish three key roles in building a sustainable future: setting jurisdiction-wide sustainability goals, developing ordinances, and levying externality fees on activities that cause people or the environment to suffer. Let's look at these roles more in depth.

SETTING GOALS

To reverse negative global environmental trends, we need both a vision of a sustainable future and specific goals to guide us. The private and non-profit sectors can help us execute this vision, but government agencies are best suited to facilitate the discussion and codify the goals, which must include: a stable climate, healthy forests, healthy waterways, thriving wildlife populations, halting mass extinction, and decarbonizing our economy.

States that have made the most progress decarbonizing their economies have set the most ambitious goals. In 2006, California passed AB 32, the Global Warming Solutions Act, which required the state to reduce greenhouse gas emissions to 1990 levels by 2020.[165] This set off a flurry of activity as the California Air Resources Board convened with over a dozen other state agencies and held public hearings to determine how every sector that burns fossil fuels can contribute to the solution. Following up on the 2006 legislation, in 2016, the California Legislature passed SB 32, which codified a 2030 greenhouse gas emissions reduction target of 40% below 1990 levels.[166] After that, Governor Jerry Brown's Executive Order B-30-15 set a goal for California to further reduce emissions: 80% below 1990 levels by 2050. Figure 35 shows the progress California has made on these goals since AB 32 was passed in 2006.

FIGURE 35: CALIFORNIA'S PATH TO PROGRESS ON CLIMATE GOALS

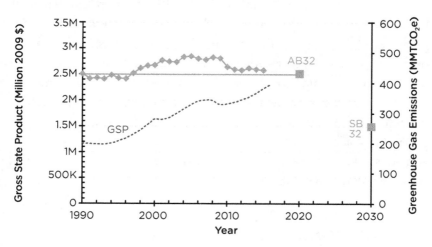

Source: California Energy Commission, using data from the California Air Resources Board's 2000-2015 inventory

Since 2006, California's greenhouse gas emissions fell, even while the state's Gross State Product grew. This shows that decoupling economic output and pollution, as economist Raworth advocates for, is possible. Achieving California's SB 32 goal of a 40% reduction below 1990 levels and the B-30-15 executive order goal of 80% below 1990 levels will be challenging, but given the commitment to innovation and coordinated efforts across multiple sectors to realize these goals, the goals should be achievable.

When governments set ambitious sustainability goals, they lay the foundation for the building blocks of a sustainable future. Sid Voorakkara, the former Deputy Director of External Affairs in the California Governor's Office of Business and Economy Development, explained how government goals bolster job creation: "When the state creates long-term goals, you attract people who want to do this kind of work. There's a gravitational pull of innovators and people who want to be part of the workforce."[167]

Goal-setting provides the regulatory certainty for the private sector to invest, because they know there will be demand for certain goods and services—thereby commencing a chain reaction. This investment is then used to create jobs that will work in support of the state's codified goals.

CREATING ORDINANCES

At the municipal level, government also plays important roles: to set goals, encourage private-sector investment in meaningful jobs, and write laws and ordinances. The latter is a tool for local governments to increase public good. For example, after establishing a deconstruction ordinance requiring houses older than 1916 to be salvaged instead of demolished and providing grant funding of $2,500 per project for one year, the City of Portland saw meaningful results in the local economy. Building materials were reclaimed, small businesses were started, and jobs were created.

Without the encouragement of the deconstruction ordinance and incentives, many people would have missed out on the opportunity to sell salvageable construction materials or receive a tax credit for donating these materials. Because of Portland's deconstruction ordinance,

TABLE 5: BREAKING THROUGH BARRIERS
FOR 30 GREAT PIVOT PROJECTS

NO.	PROJECT
1	Zero Net Energy Retrofits—Single Family Homes
2	Zero Net Energy Retrofits—Multi-Family Dwellings
3	Solar Emergency Microgrid Retrofits at Hospitals and Municipal Emergency Response Centers
4	Commercial Solar Siting Surveys for Counties
5	Zero Net Energy Retrofits—Commercial
6	Part-Time Embedded Sustainability Project Manager
7	Development of Mass Transit
8	Transportation Management Associations
9	Development of Safe Bicycling Infrastructure
10	Mobility-as-a-Service Apps
11	Electric Vehicle Charging Infrastructure
12	Designing Walkable Communities
13	Local Government Waste Prevention Coordinators
14	Building Deconstruction

BARRIER(S)	SOLUTION(S) TO BREAK THROUGH	WHO CAN HELP
Learning curve for the homeowner	State incentives for contractors to provide bundled offering of energy efficiency retrofits, solar installation, and financing	State energy commission
Split incentive	Pay As You Save funding program	Utilities
Competing priorities	Require ZNE retrofits over long-term time horizon	State
Lack of funding	Government funding	State or county
Split incentive	State incentives for contractors to provide bundled offering of energy efficiency retrofits, solar installation, and financing	State energy commission
Lack of funding for contract labor	Upfront funding repaid with shared savings	Finance industry
Lack of funding	Green bonds	State
Lack of funding	Charge for parking downtown	Municipality
Lack of funding	Funding	Municipality
Apps need further development	Funding	Private equity
Lack of funding	Goal setting for EVs and EVCI, public-private partnerships	State
Lack of information	Workshop for municipal transportation staff highlighting success stories	Non-profits
Lack of funding	Waste prevention goals and landfill tipping fee to provide funding	County and state government
Demolition is less expensive	Deconstruction ordinance and grants	Municipalities

THE GREAT PIVOT

NO.	PROJECT
15	Tool Lending Library + Repair Cafe + Maker Space
16	Upcycling Dead or Diseased Trees
17	Artistic Upcycling and Salvage
18	Regional Recycling Market Development Managers
19	Reverse Catering
20	Community Kitchens
21	Ugly Produce Distribution
22	Business Services Supporting Small, Organic Farms
23	Carbon Farming
24	Restoring Healthy Forests
25	Construction Products and Furniture
26	Restoring Healthy Waterways
27	Wildlife Defense
28	Wildlife Restoration
29	Wildlife Overpasses and Underpasses
30	Rewilding for Habitat Restoration

BARRIER(S)	SOLUTION(S) TO BREAK THROUGH	WHO CAN HELP
Lack of space	Public libraries provide small space initially with room to grow	Local government
Lack of equipment, markets	Financing, marketing	Financial sector
Unaware of opportunity to increase revenue	Online workshops to share best practices	Non-profits
Lack of recycling goals, lack of funding	Set goals for regional recycling and landfill tipping fee to provide funding	County and state government
Low cost of food disposal	Financial disincentive for food waste disposal	Municipality
Lack of funding	Funding	Philanthropy and government
Lack of knowledge	Marketing	Finance industry
Expertise in agriculture not business	Business services support	Non-profits
Lack of funding	Cap and trade fee	State
Lack of funding	Cap and trade fee	State and federal government
Insufficient entrepreneurs	Business incubators	State and federal government
Lack of funding	Parcel tax	State
Lack of funding	Funding	State
Lack of funding	Funding	State
Lack of funding	Funding	State
Lack of funding	Funding	State and federal government

building materials were handled in a way that reduced dispersal of hazardous materials like lead, asbestos, and mercury into neighborhoods.

THOSE BEST EQUIPPED TO LEAD SUSTAINABLE CHANGE

With the disruption toolkit of new metrics, new goals, local ordinances, and externality fees in mind—as well as the strengths that the private, non-profit, and public sectors bring to the table—let's take a look at how we can jumpstart the 30 meaningful job projects the Great Pivot describes. Listed are major barriers to each project, key solutions for overcoming each barrier, and an agent in the best position to help.

Work to overcome these barriers happens synergistically across different sectors, with each sector contributing according to their strengths and resources. A municipality creates an ordinance that encourages the private sector. The financial sector provides funding to grow a market. Non-profits facilitate discussions among municipal managers. The state provides funding for restoration work run by non-profits and carried out by contractors. Working together, each sector brings to the table what they do best and complements each other.

Tackling a challenge as big as climate change will require each of us interested in doing this work to bring our "A game" and work well as a team. The November 2018 UN IPCC report with 91 authors and review editors from 40 countries urged strengthening the global response to the threat of climate change. Focused on the hope of keeping the average global warming to 1.5°C above pre-industrial levels, the report urges "rapid and far-reaching" transitions in land, energy, industry, buildings, transport, and cities to reduce fossil fuel combustion and sequester carbon. "The good news is that some of the kinds of actions that would be needed to limit global warming to 1.5°C are already underway around the world, but they would need to accelerate," said Valerie Masson-Delmotte, Co-Chair of Working Group I, the IPCC working group assessing the physical science of climate change.[168]

The still-living researchers who modeled World3 in the early 1970s might be interested to learn that the IPCC makes pains to point out in the IPCC report, like in Graham Turner's 2014 update, that humanity is only doomed if we do not change anything and continue with business as usual. The IPCC explained that if we make the bold choice of bankrupt-ing the fossil-fuel industry and share the burden of transition equally, most humans can live in a world better than the one we have now.[169]

CHAPTER HIGHLIGHTS

- Business as usual is leading civilization straight toward a cliff. To change course, we need different goals and metrics.
- The World Happiness Report measures five factors besides gross domestic product to determine citizens' level of happiness in various countries. Those five factors include social support, healthy life expectancy at birth, freedom to make life choices, generosity, and perceptions of corruption in government and business.
- Government agencies can help steer society in a more sus-tainable direction by setting goals and creating ordinances.
- Decoupling economic growth from GDP involves tran-sitioning from fossil fuels to renewable energy, creating a circular flow of materials, dematerializing, and radically reducing waste.

10

INVESTING IN THE WORLD WE WANT

*"I believe in the power of nurture capital to drive positive change.
We should see capital as a tool, not the master."*
—Marco Vangelisti, cofounder of Slow Money

Merrilee Olson, the founder of Preserve Farm Kitchens, waited outside a conference room in San Francisco to make her funding pitch to peer-to-peer lending organization Northern California Slow Money. Her company needed its third round of funding to expand into a new location. She presented to Slow Money's Slow Opportunities for Investing Locally (SOIL) group where twelve potential investors heard two pitches before deciding on their level of interest in the food, farm, or fiber entrepreneurs and their companies.

In less than five minutes, Merrilee explained that Preserve Farm Kitchens takes fruits and vegetables that farmers can't sell, either because of overproduction or because the crop is cosmetically unacceptable, and turns them into jams, sauces, and condiments. She outlined what it would take for her company to reach the next developmental stage, and that she was looking for $100,000 of funding. Investors asked her questions for twenty minutes and went around the table giving her feedback.

The next day, her assigned champion called to tell her how the presentation went. Merrilee then had follow-up meetings with three potential investors. Together, they discussed their interest in her small business and set the loan amount and terms in the investment instrument that would best serve all involved.

Of the pitching experience, Olson said, "Getting to present before the Slow Money audience is really a pleasure. You get so much positive feedback. You don't have to worry about your audience because you know you're in front of a self-selecting group of investors who are committed, who are also very passionate."

In the Northern California Slow Money chapter in the South Bay, deal sizes range from $5,000 to $15,000. The San Francisco/Berkeley chapter makes deals that may be up to a few hundred thousand dollars. The size depends on what the entrepreneur needs and are usually for an expense that will help the entrepreneur take their business to the next level; some requests are for purchasing a piece of equipment and others to construct a kitchen. The rate of return for Slow Money deals often range from 5–7%. Some investors will accept an interest rate below 5% if the venture involves a compelling social justice angle.

Almost any investment of this type triggers securities laws, including both federal laws and state laws, so these pitches at Slow Money are carefully designed to be in compliance. In particular, securities laws distinguish between private offerings, where there is a personal connection between the person raising capital and each of the potential investors, and public offerings, which can be openly advertised. Public offerings can reach a much-wider investor base, including non-wealthy investors, but involve more work, because they typically need to go through a regulatory review process. So, these pitches are strictly private affairs, not open to the public.

Building relationships with social entrepreneurs appeals to Slow Money investors like Roy Kornbluh. He said he was visiting a Farmer's Market in Olympia, Washington, when he saw a sign that urged visitors, *Don't buy food from strangers.* Kornbluh said, "I want to know where my food and my fiber are coming from. Slow Money gives us the chance to create the society we want with our pocketbooks."

Slow Money invests in small food, farms, and fiber (organic cotton and wool) social enterprises around the country as a way to build local economies, but peer-to-peer lending is just one support mechanism that provides advice and financing to help transition our society to a more sustainable future.

People who are brave enough to want to strike out on their own and start a business deserve financial and emotional support to do so. They need funding and the security to know they won't lose their home or incur crippling debt if their business fails or if they encounter unforeseen obstacles. Starting your own business is scary, and our country must make it easier for aspiring entrepreneurs to do so.

Whether someone wants to pursue meaningful work in the private, government, or non-profit sector, there are a number of projects that could serve as inspiration. Solar energy has seen serious investments lately, but what about other projects that will help build a sustainable food system or a circular economy? This chapter describes myriad financial tools that have been used to support those who sought to make their dream of non-mainstream work a reality.

Like Roy Kornbluh, some members of the financial sector want to do more than simply make money: they want to do their part to address the crises we are facing. Ahead of the 2018 UN conference on climate change, global investors managing $32 trillion (with a "t") in assets called on governments to accelerate efforts to combat it. Investment firms such as UBS Asset Management and the New York State Common Retirement Fund signed the "2018 Global Investor Statement to Governments on Climate Change," demanding urgent action. While federal, state, and local governments have important roles to play in setting the goals, regulatory frameworks, and financial incentives that will guide the transition from a fossil fuel-intensive to a low-carbon society, we now need the muscle of the financial sector to scale up the process.

The financial sector is like a wild horse: if we could simply harness it, we could steer its investments toward advanced energy communities, carbon-free mobility options, and a circular economy, thus galloping into a sustainable future, rather than moving at the current meandering trot. The financial sector's abundant resources could be put in service of an ambitious goal like a clean energy economy. For example, the New York State Common Retirement Fund alone oversees $207 billion in assets.

A study published in 2018 found that if the New York State Common Retirement Fund had divested its fossil fuel stocks in 2008, they would have been $22 billion richer in 2018—$19,820 more for each of the fund's 1.1 million members. Toby Heaps, CEO of Corporate Knights, the investment and media research company that conducted the study, remarked, "Our analysis reflects that the energy transition is happening faster than investors expected. Those responsible for safeguarding pension assets now need to ask themselves if they believe this transition is over or just beginning."[171]

To build a clean energy future, it's not enough to screen out fossil fuel industry investments from a portfolio. Rather, retirement funds need more sustainability projects to choose from and invest in. The problem is that there are not enough business plans, funds, or green bonds to meet the need. Meanwhile, many fledgling yet promising projects in the private sector could deliver a competitive return on investment if they were fostered.

The Great Pivot describes 30 projects, some of which are well suited for the private sector, some for the public sector, and some that work best in the non-profit sector. The private sector is a good match for Great Pivot projects that offer a competitive or above-competitive return on investment. Funding from the financial sector for private sector projects would be like installing a battery in an electric car: without the battery, the electric car can't go anywhere, but once it has a power source, the electric car can accelerate quickly.

Great Pivot projects for the private sector include:

(1) Zero Net Energy Retrofits—Single-Family Homes
(3) Solar Emergency Microgrid Retrofits at Hospitals and Municipal Emergency Response Centers
(5) Zero Net Energy Retrofits—Commercial
(6) Part-Time Embedded Sustainability Project Manager
(10) Mobility-as-a-Service Apps
(14) Building Deconstruction
(16) Upcycling Dead or Diseased Trees
(21) Ugly Produce Distribution
(25) Construction Products and Furniture

These nine private sector projects will provide a financial return on investment, but the rest provide a different type of return, as they will yield important social and environmental benefits for the public:

(2)   Zero Net Energy Retrofits—Multi-Family Dwellings
(4)   Commercial Solar Siting Surveys for Counties
(7)   Development of Mass Transit
(9)   Development of Safe Bicycling Infrastructure
(11)   Electric Vehicle Charging Infrastructure
(13)   Local Government Waste Prevention Coordinators
(15)   Tool Lending Library + Repair Cafe + Maker Space
(18)   Regional Recycling Market Development Managers
(24)   Restoring Healthy Forests
(26)   Restoring Healthy Waterways
(28)   Wildlife Restoration
(29)   Wildlife Overpasses and Underpasses
(30)   Rewilding for Habitat Restoration

An important point to note is that these public sector projects may be funded or led by a government agency but rely on critical support from the private sector and/or non-profit sector. Commercial solar siting surveys may be funded by county governments but conducted by non-profits. Community engagement for the development of safe bicycling infrastructure will be conducted by municipal transportation departments, but private contractors will install protected bike lanes. Funding for restoring healthy waterways may be pieced together by elected representatives at the federal and state level, but the work will be done by private sector contractors and coordinated by non-profits.

The last group of Great Pivot projects will be best served by the non-profit sector. These organizations are lean and nimble, and such qualities provide the space to experiment with innovative projects and show us what is possible, all while serving the public good. Great Pivot projects well-aligned with the non-profit sector include:

(8)   Transportation Management Associations

(12) Designing Walkable Communities

(17) Artistic Upcycling and Salvage

(19) Reverse Catering

(20) Community Kitchens

(22) Business Services Supporting Small, Organic Farms

(23) Carbon Farming

(27) Wildlife Defense

In order to create jobs in these 30 areas, these projects require funding. To figure out where to source that funding, let's look at how similar projects found theirs.

PRIVATE SECTOR FUNDING

A wide range of potential investors have the means to fund job creation in social enterprises, which are organizations that address a basic unmet need or solve a social or environmental problem through a market-driven approach. For aspiring social entrepreneurs who require funding to grow their enterprise to the next level, the first thought may be to seek out professional investors like angel investors, venture capitalists, and private equity managers. However, with new legislation relaxing the regulatory burden for securities offerings and evolving crowdfunding tools, more non-professional investors can help fund green projects.

From the point of view of the job-creating entrepreneur, funding is inherent to the development stage. On the journey from concept to maturity, different sources offer various types of funding. Figure 36 illustrates the conventional wisdom about the typical sequence of funding sources for tech startups as they work through various stages toward their product launch, at which point their revenues will start to exceed their expenses.

FIGURE 36: INVESTMENT LIFECYCLE STAGES FOR STARTUPS

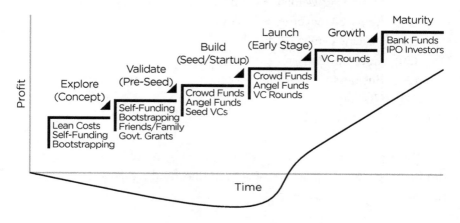

Source: Nicholas Petit

This graphic applies mainly to companies that can scale quickly, because they make software or products. Some of the elements of this graphic also apply to social enterprises that provide services: in the concept exploration phase, companies can bootstrap and focus on lean costs in order to self-fund; during the validation phase, they can branch out to friends and family but still keep costs lean. Then, as the social enterprise prepares to grow, the enterprise will need more funding to scale up.

The build, launch, and growth phases for social enterprises, however, require different sources of funding than for software and products. Let's look at the nine private sector Great Pivot projects, for within these projects lie proofs of concept that reveal where funding can be found.

Zero Net Energy Retrofits—Single-Family Homes
Funding need: worker salaries

A small business that does ZNE retrofits for single-family homes needs funding to bridge the time gap between when the retrofit work is done and when funding comes through. When combined, utility rebates, government tax credits, a home equity line of credit, and other funding options can help homeowners pay for their energy efficiency retrofit and renewable energy systems. This funding is available at various times throughout the installation, including before work begins or long after it's done. Meanwhile, workers need a regular paycheck, and various organizations can help bridge the funding gap for ZNE retrofit business managers.

Pioneer Valley Photovoltaics (PV Squared) is a worker-owned cooperative that installs solar in Connecticut and Massachusetts. At one point, the company had a cash flow problem due to the nature of their solar installation business. Most of their clients use state-sponsored solar rebates to pay for a chunk of their installation, but these rebates don't become available until after the system has been installed and is operating. To bridge the gap and pay their staff before state photovoltaic rebate checks arrived, PV Squared secured a line of credit from Cooperative Fund of New England, a Community Development Loan Fund (CDLF). CDLFs facilitate socially responsible investing in cooperatives, community-oriented non-profits and worker-owned businesses in New England.

Kim Pinkham, a worker-owner at PV Squared, explained that her business's line of credit from the community loan fund helped them convert more homes and businesses in Connecticut and Massachusetts to solar energy. "The line of credit from the Cooperative Fund helped create three jobs and retain twelve," said Pinkham. "We use the line of credit to cover payroll during installations and to make up any expenses that aren't covered by our clients' payments. It gives us breathing room we wouldn't otherwise have."[172]

With more support from CDLFs, the U.S. could scale up renewable energy and energy efficiency retrofits in single-family homes.

Solar Emergency Microgrid Retrofits at Hospitals and Municipal Emergency Response Centers
Funding need: capital project equipment and labor

The installation of solar emergency microgrids at hospitals and municipal emergency response centers require large financial outlays. In addition to equipment costs, the projects involve the planning, design, and installation of energy-efficient equipment, renewable energy systems, electric vehicle charging stations, energy storage, and monitoring, communications, and controls—all of which demand long-term capital and strategic planning. Forward-thinking hospitals and municipal response centers realize that if they install a solar emergency microgrid, they will be able to serve their community even when the electric grid is down. The monitoring, communications, and controls equipment in a solar emergency microgrid allows the system to "island" from the grid and still run critical loads.

In 2017, Kaiser Permanente installed a solar emergency microgrid at its Richmond, California, facility. The project was made possible by a $4.75 million grant from the California Energy Commission (CEC). Kaiser contributed $2 million of "match" funds by doing an LED lighting retrofit, installing a dozen electric vehicle chargers, donating staff engineers' time to plan the project, and securing material and equipment discounts from project suppliers.

This microgrid project is important to Kaiser, because they understand the health implications of climate change. Kaiser's mission focuses on total health, which they describe as having "multiple, interrelated dimensions [including] the physical, emotional, and spiritual health of every individual, supported and sustained by the health of our total environment."[173] In this spirit, Kaiser is reducing their carbon footprint by installing solar and energy storage.

Kaiser's Richmond facility microgrid includes 250kW of solar photovoltaics (which sits atop Kaiser's parking garages), smart inverters, 1MW of batteries, and a microgrid controller. The $4.75 million of funding came from a surcharge on electricity ratepayers from Pacific Gas & Electric, San Diego Gas & Electric, and Southern California Edison. This surcharge goes into a fund called the Energy Program Investment Charge (EPIC), and the California Energy Commission's EPIC program distributes $162 million annually to "support investments in clean energy technologies that provide benefits to electricity ratepayers."[174]

Organizations in states that do not have a fund like EPIC will have to pinpoint alternative means for financing projects like this. Fortunately, capital groups are stepping up to the plate. Innovative private equity firms perform a calculus by piecing together variables such as funding available from grants, rebates, and tax credits, reducing demand through energy efficiency projects, changing to a lower electricity rate, and using batteries to store electricity for the time of day when electricity is more expensive in order to create a profitable Power Purchase Agreement. Capital groups calculate the minimum amount that can be spent on a microgrid project to save the maximum amount of money on the bill. Then they structure a deal that helps organizations with insufficient capital availability install a microgrid that results in them paying the same amount per month for their energy bill as they did before the project.

Solar Emergency Microgrid at Santa Rita Union School District

Investment and operating platform Generate Capital structured a $5 million finance package for a ground-breaking combination of microgrid technologies at schools in the Santa Rita Unified School District in Salinas, California. With this deal, the school district now has solar emergency microgrids at six schools that can keep the schools running during power outages, and they are paying the same price for electricity per kilowatt hour as they previously did for grid power.

At each of the schools, project partners installed energy-efficient lighting, 115kW to 264kW of solar photovoltaic panels, wireless thermostat controls, 80kWh of batteries (to provide seven hours of energy storage), bi-directional inverters, and monitoring, communications, and control equipment to manage the flow of electrons. This combination of technologies allows the systems to shift loads so the schools can buy electricity when it's cheaper and use stored electricity when it's more expensive.

A combination of grant funding and load shifting made the project financially viable. Generate Capital and SRUSD created a Power Purchase Agreement that pieced together a Self-Generation Incentive Program rebate from the California Energy Commission for energy storage equipment and Prop 39 funding from the state for schools to make their lighting and HVAC systems more energy efficient, and they also worked with Pacific Gas & Electric to switch to a lower utility rate. The deal was enhanced by net metering, which allows SRUSD to generate solar power in the summer when school is not in session and bank peak kilowatt-hour prices to help fund the energy storage equipment. This project provides a community resource of uninterruptible power in case of a prolonged power outage and offers students a living classroom of clean power technologies.

Zero Net Energy Retrofits—Commercial
Funding need: equipment and labor

One tool to finance ZNE commercial building retrofits is a Property Assessed Clean Energy (PACE) loan. As more and more states and local governments pass PACE legislation, allowing building owners to roll costs of energy efficiency and solar photovoltaic projects into their building's property taxes, more people will have access to a financial tool that makes these types of upgrades easier. PACE loans do not count as debt on a commercial building owner's balance sheet, which is advantageous to the building owner, and are repaid over the selected term, from 5–25 years.

Mynt Systems, the Santa Cruz-based commercial building contractor mentioned in chapter four, works with building owners to design, engineer, and implement commercial building ZNE retrofits using PACE loans. One building Mynt worked on, a 40,000 square foot office building in Oakland, California, received upgrades to its HVAC system, to LED lighting, solar panels, window film, and sensors and controls throughout. Given how energy intensive the coworking robotics lab is on one floor, the building is nearly ZNE. The building owner took out a fifteen-year PACE loan at 5.85%, which will pay for itself in just under five years. The tenants will see their rents rise post-retrofit, but the savings in utility payments will be greater than the rent increase. The building owner, on the other hand, now has a more valuable building and can cover PACE loan payments by charging higher rents.

As paybacks become increasingly attractive though, more commercial building owners are self-funding ZNE retrofits. Based on his experience, Mynt Systems CEO Derek Hansen said:

> "I believe that the greatest source of funding for these type of projects throughout the energy retrofit space will always be

the landlords, developers, and property ownership groups. It has been somewhat effective to look for outside private funding, but once the landlord sees the performance metrics (ROI), they tend to circumvent the need for external funding."

Part-Time Embedded Sustainability Project Manager
Funding need: gap financing for labor

Small and medium-sized businesses (SMB) don't usually hire sustainability managers, yet many would benefit from having one, even part time. For SMBs that spend hundreds of thousands of dollars per year on operating costs such as energy bills, feedstock, supplies, garbage bills, and water bills, being able to hire a part-time sustainability manager to implement operational efficiency projects will save the business money. With limited cash flow, SMBs need a way to fund a contract-based, part-time sustainability manager until savings start to accrue.

One worker-owned cooperative in Boston, Cooperative Energy Recycling and Organics (CERO), developed an innovative funding solution to bridge the funding gap between the time their employees need to be paid and when their clients realize cost savings on their garbage bill. CERO creates jobs in the low-income Boston neighborhoods of Roxbury, Dorchester, and East Boston, picking up organic materials for composting from businesses, which lowers their garbage bills.

With the help of Cutting Edge Capital, an advisory firm in Oakland, California (and its sister law firm, Cutting Edge Counsel), CERO created a Direct Public Offering (DPO) to raise funds to grow their business. A DPO refers to a public offering of securities to both accredited and community (non-accredited or non-wealthy) investors. Businesses and non-profits can use DPOs to market and advertise their offering publicly in newspapers, magazines, at public events and private meetings, and on the internet and through social media channels. While DPOs are sometimes referred to as "do-it-yourself IPOs" (because no intermediary is involved) and "securities-based crowdfunding," securities law can be a tricky thing. Having legal advice from experts like Cutting Edge can be critical to success.

CERO set a targeted fundraising goal of $340,000 with a minimum investment at $2,500. They offered a 4% dividend over a five-year term. The five CERO employees conducted a public outreach campaign and received community investment from throughout Boston with strong support coming from the Dorchester neighborhood where the cooperative is located. Ultimately, the DPO received investments from 83 investors.[175]

Mobility-as-a-Service Apps

Funding need: labor for application programming and marketing

Ideally, Mobility-as-a-Service apps help people plan the fastest route for local trips using a combination of low-carbon mobility options and then facilitate seamless payment for the entire combination of options used. New startups in the MaaS space need funding for programmers to develop the app and sales and marketing people to develop a client base.

Impact investors offer a good fit for funding MaaS apps. Silicon Valley Social Venture (SV2) describes itself as venture philanthropy, providing grants as well as making impact investments in social enterprises. They seek out angel- or seed-stage investments in social enterprises hoping to raise up to $1.5 million but are open to those seeking up to $3 million. In 2017–2018, SV2 distributed $500,000 in cash grants and investments and $226,000 in experiential grantmaking and impact investing while raising funds from SV2 partners, the board of directors, and institutional foundations.

One of SV2's 2015 impact investees was PastureMap, a mobile app that allows cattle ranchers to manage their grasslands more efficiently and profitably while sequestering carbon and reducing CO_2 emissions. PastureMap allows ranchers to see their entire ranch on a map, pinpoint where their herd has been, and plan upcoming herd drives based on weather and drought. It tracks cumulative forage grazed in each pasture, allows ranchers to upload and view photos they take in the field (to track pasture inventory), and exports grazing records to Excel so the whole team can stay on the same page. The app is used by 9,000 ranchers in 40 countries.

Investing networks like SV2 may be comprised of angel investors, venture capitalists, foundations, and family offices with the intent to fund early stage social enterprises. Another example of an investing network is Social Venture Circle. Established in 1987, Social Venture Circle has facilitated $200 million of investments in over 330 social ventures that address environmental, education, health, and community challenges. Some of their early stage investments include ZipCar and HonestTea.

Building Deconstruction

Funding need: tools, a truck, certification, licensing, and insurance

David Greenhill was listening to the radio one day in 2016 when he heard about the City of Portland's new deconstruction ordinance. The city was going to start requiring deconstruction of houses more than 100 years old, instead of being demolished, and the construction materials would then be sold or donated. At the time, Greenhill was working in graphic design and wanted to do something different. Deconstruction sounded like a worthwhile endeavor.

Instead of taking out a loan from a bank, which would have cost him $100 in interest per month, Greenhill kept his startup costs low and bootstrapped. Using funds from savings, he bought the bare minimum to get up and running: $2,000 for an old pick-up truck, $3,000 worth of tools, a few thousand dollars for certification, a commercial driver's license, and workers' compensation insurance. All in all, he spent about $10,000 to get his new business, Good Wood, up and running.

When he brought on workers—friends and friends of friends at first—the agreement was informal and offered an hourly wage. Cash flow was a challenge at the beginning. The company would receive a deposit for a deconstruction job and Greenhill would pay wages out of that. He built the business up as his crew deconstructed more houses and then sold the lumber from the jobs out of Greenhill's back yard.

Good Wood now has a 1,500-square-foot warehouse and sells about $10,000 per month in salvage wood. In November 2018, they sold a banner $23,000 worth of salvage wood. "You have to hustle, but I didn't have to borrow to build this business." He asserted that the City of Portland creating an ordinance requiring homes before 1916 to be deconstructed and giving away $2,500 grants for a year were key to him being able to develop his business. Greenhill is proud that he created eight jobs out of his initial $10,000 investment. "We created something out of nothing. It's awesome."

Upcycling Dead or Diseased Trees
Funding need: vehicles, equipment, and tools

Tree removal companies use chainsaws, rigging, and a chip truck to turn a dead or diseased tree into firewood, mulch, and sawdust, then haul it off-site. This labor-intensive process downcycles a tree into low-value products and waste. For tree removal companies that aspire to upcycle unwanted trees into lumber, a few pieces of equipment can be purchased or rented. Depending on the types of jobs and the location of the target trees, an urban lumber company may want to have access to a Tree-Mek, a crane, a grapple truck, and/or a portable sawmill.

A Tree-Mek is a truck with a hydraulic arm that can telescope up to 120' to grab a section of a tree, cut the section right below where the grabber attaches, then gently deposit the tree section on the ground. For larger tree removal companies, Tree-Meks and cranes can cost hundreds of thousands of dollars, an investment that would yield an attractive return on investment for large-volume jobs. For smaller companies that do this kind of work less frequently, these pieces of equipment can often be rented by the day or for a few hours and come with trained, certified operators. For jobs extracting large trees that could potentially yield 10–20 tons of lumber, cranes are often used. Dave Hunzicker with Urban Hardwoods in Seattle explained, "Eighty percent of the urban trees we get have a crane involved to extract them."

Others find that grapple trucks are an invaluable piece of equipment. Peter Gruenwoldt, President of Seattle Tree Care, took out a five-year commercial bank loan to purchase one. "You can buy them used for about $125,000 or new, top of the line for $220,000 plus tax. It's an amazing tool."

Urban lumber advocate Dave Barmon with Epilogue in Portland said that $45,000 invested in a portable sawmill is money well spent, given that his company is able to turn more removed urban trees into lumber. Two of his workers can shift the 26' portable sawmill from the work truck to a client's front yard and use it on-site, obviating the expense of hauling logs to the nearest sawmill and allowing the company to create a higher-value product. This is another piece of equipment for which businesses take out commercial loans. He prefers milling lumber at his sawmill, as doing so is more efficient, but a portable sawmill is another option. Barmon cautions aspiring entrepreneurs interested in moving into urban lumber to keep in mind that removing trees and milling wood can be extremely dangerous and that having proper training and insurance is crucial before providing such services.

Ugly Produce Distribution
Funding need: labor, vehicles, and warehouse space

Imperfect Produce uses clever marketing to create demand for ugly produce and thus help fix the massive societal problem of food waste. Produce distribution is a capital-intensive business though, and Imperfect has needed to raise multiple rounds of funding to pay for trucks, warehouse space, labor for logistics, marketing, and other aspects of their operations. Imperfect has been able to attract private equity funding in part because of the sweet niche the company occupies: they are rescuing produce that would otherwise be thrown away, their work helps boost farmers' revenues, and they deliver produce to customers' doorsteps at prices cheaper than grocery stores charge.

As of June 2018, Imperfect had raised $10 million in equity funding and was looking to raise tens of millions more in the near future. To date, private equity firms such as Maveron, Flybridge Capital Partners, Shasta Ventures, Correlation Ventures, Ranch Ventures, Fresh Source Capital, and Corigin Ventures have contributed funding.

Construction Products and Furniture
Funding need: work space, marketing, and tools

A small woodworking shop that wants to turn salvage lumber into furniture would need work space, marketing help to drive demand for the products, and additional tools. A crowdfunding effort would allow a small business like this to expand. One key to successful crowdfunding is being able to tell a compelling story, preferably through a well-produced video. Taking trees damaged by bark beetle infestations and turning the salvage wood into pieces of furniture, all while highlighting the beauty of the material (as aforementioned Ghost River Furniture does), offers a trash-to-treasure story that could easily attract potential funders. A small business of this type would not need a lot of funding to take their business to the next level.

One crowdfunding platform for entrepreneurs who need microloans of $10,000 or less is Kiva. Ryan Schmidt, proprietor of Mitty's Metal Art, a blacksmith shop and gallery in Cumberland Gap, Tennessee, used Kiva to raise $10,000 from 177 lenders. This allowed him to purchase additional tools and expand his metal art studio to include a storefront that could accommodate his creations and those of other local artisans and craftspeople. He also hired a local marketing expert to design his logo and brand and overhaul his website. These efforts turned his shop into a tourist attraction for their small, historic town.

While Schmidt is repaying his zero-interest Kiva loan over 36 months, other private-sector Great Pivot projects pay their investors competitive or above-competitive returns on their investments. PV Squared took a line of credit from a Community Development Financial Institution. PastureMap received impact investment funding from a social venture fund. Imperfect Produce secured private equity funding from several venture capital firms. Many financial tools are available for social entrepreneurs who have business plans with a healthy return on investment, and while the private sector is a good fit for many Great Pivot projects, we also need the participation of the public sector—help that will support projects yielding social and environmental benefits.

PUBLIC SECTOR FUNDING

In 2015, the City of Watsonville, California, started levying a Carbon Impact Fee on all new residential and non-residential development projects, as well as major additions and alterations of existing buildings. The purpose of the fee is to incentivize energy efficiency in buildings. Projects that incorporate energy-efficient measures receive a partial refund of the fee from the city, and ZNE projects receive the entire fee back.

Watsonville has a large number of multi-family dwellings that the city wanted to make more energy efficient. A split incentive stood in the way of progress, wherein low-income tenants could not afford to pay for energy efficient upgrades but would benefit from a lower energy bill each month, and landlords do not pay the energy bills, meaning they do not have an incentive to invest in energy efficient upgrades in the dwellings. Thus, the city created a Carbon Impact Fee to incentivize building efficiency upgrades.

Government plays a key role in resolving vast, complex problems like climate change. To raise society's response to a level that matches the severity of the threat of climate change, the public sector must employ the financial tools at its disposal, including taxes, fees, and bonds. All three generate revenues that governmental agencies can use to create meaningful jobs for developing clean local energy systems, establishing low-carbon mobility infrastructure, building a circular economy, and restoring natural systems.

Taxes

While fees are voluntary payments to the government for the special services rendered by it in the public's interest, taxes are compulsory payments the government levies on its citizens and businesses. Here are several different types of taxes, some of which have been used to fund sustainability projects.

- Income tax
- Corporate tax
- Sales tax
- Property tax

- Parcel tax
- Tariffs
- Estate tax
- Hotel tax
- Utility tax

Utility taxes fund energy efficiency rebates for utility customers, while parcel taxes have been used to fund environmental restoration work. As an example, a 2016 ballot referendum in the San Francisco Bay Area, called Measure AA Save the Bay, asked voters in the nine counties that touch the San Francisco Bay if they would like to restore the health of the Bay through a $12-per-parcel tax each year for twenty years. To the delight of environmentalists in the Bay Area, 70% of voters approved the referendum.

Sam Schuchat with the San Francisco Bay Restoration Authority announced, "This is the first time that a region has voted to tax itself for wetland restoration. It's really highly significant."

Each spring, $25 million in parcel tax money flows to the San Francisco Bay Restoration Authority, a nine-county agency that oversees bay conservation projects, and the Authority then allocates funding for various projects. In 2018, projects funded by Measure AA included:

1. $8 million for the South Bay Salt Pond Restoration Project (San Mateo, Alameda, Santa Clara counties)
2. $4 million for the South San Francisco Bay Shoreline Project (Santa Clara County)
3. $2 million for the Montezuma Wetlands (Solano County)
4. $1 million for the Deer Island Wetlands (Marin County)
5. $1 million for the San Leandro Treatment Wetland (Alameda County)
6. $3 million for the North Bay wetland restoration (Sonoma, Marin counties)
7. $150,000 for Lower Sonoma Creek (Sonoma County)
8. $450,000 for Encinal Dune (Alameda County)
9. and, pending approval, $5 million for the India Basin remediation (San Francisco)

With all the efforts to restore the Bay, area residents have watched the 300-acre salt pond in the South Bay revert to wetlands.

David Lewis, Executive Director of the non-profit Save the Bay, pledges to parcel tax payers that their twelve dollars is going to "great stuff."

"First comes the return of the water," he said, "followed by the return of everything else. Plants, fish, birds, seals, sea lions, and sharks. The brown will turn to green. Soon the bay water will start coming in and out twice a day, just like the moon makes it do."

John Bourgeois, manager of the South Bay Salt Pond Restoration Project (another partner on the multi-decadal San Francisco Bay restoration project) took it a step further, explaining, "This is the largest wetland restoration project on the West Coast." Drawing a mental picture of the work ahead, Bourgeois invited us to picture 50,000 dump trucks bringing enough dirt into the park to shore up two miles of old earthen barriers. He explained that this preparation work amounts to "setting the table for inviting Mother Nature to come back and do her work."

Then, one happy morning in the near future, a backhoe will knock down a section of the old berm, and the San Francisco Bay will rush back in, followed by countless fish, birds, and other critters who know progress when they see it. Then the marsh plants will convert the tract from salt-white to marsh-green.[176]

Twelve dollars per parcel may not seem like a lot of money, but when aggregated, it makes a big difference in restoring the health of the San Francisco Bay after more than a century of pollution and development. This funding also makes a difference to the people both doing meaningful work to restore habitat for endangered species and creating buffers against a changing climate. The Measure AA ballot initiative provided key funding for these projects.

Fees

Another ballot initiative that created a funding stream for important sustainability work happened back in 1989. Voters in Alameda County, California, were asked to approve Measure D, which called for a 75% reduction in landfill waste by 2010. At the time, only 14% of discarded materials were being diverted from landfills. A majority of voters approved Measure D, and the quasi-governmental agency StopWaste was formed in 1990, supported by surcharges levied on solid waste tipped into Alameda County landfills.

The funds from these levies were used to hire people who developed and administered waste reduction programs—and the programs worked. In 1990, two million tons of garbage headed to landfill; in 2015, only one million tons of garbage went to landfill, even though Alameda County's population was 25% larger than in 1990.

Alameda County StopWaste now employs 43 staff members who help the county's businesses, residents, and schools waste less, recycle and compost more, and use water, energy, and other resources more efficiently. Their current $30 million annual budget comes from a few different sources:

- $4.34 per ton facility fee levied on Alameda County solid waste landfilled in California
- $2.15 per ton household hazardous waste fee
- $8.23 per ton fee collected on waste disposed in Alameda County landfills
- $4.53 per ton import mitigation fee collected on all waste landfilled in Alameda County that originate out-of-county

The budget not only covers agency staff: a portion of the funding is distributed to municipalities for their staff to work on waste-reduction projects with businesses, residents, and schools. This constant stream of funding has allowed StopWaste to help the county achieve 67% waste diversion and continue driving the county toward a circular economy. The county's long-term aspirational goal is that "less than 10% of the good stuff ends up in landfill."

In a way, Alameda County StopWaste is a victim of its own success. As they work to help the county reduce the amount of waste, fees collected to do waste-reduction work continue to fall. From the perspective of building a circular economy and a sustainable future, ideally, at some point, StopWaste staff will put themselves out of a job. At this point though, there is still work to be done.

Green Bonds

Besides taxes and fees, bonds are another financial tool that can generate funding to create green jobs. Green bonds refer to a fixed-income financial instrument for raising capital used to fund environment-friendly projects. Proceeds from these bonds are earmarked for green projects and are backed by the issuer's balance sheet. The bond issuer, such as a state or municipality, raises a fixed amount of capital from investors for an established period of time (the "maturity"), repays the capital (the "principal") when the bond matures, and pays an agreed-upon amount of interest ("coupons") during that time. Green bonds are typically issued by government agencies (and are therefore a type of municipal bond), because the interest paid on these bonds may be tax-exempt for the investors.

In June 2013, the first green municipal bond was issued by Massachusetts to finance the state's Accelerated Energy Program, which aimed to reduce energy consumption by 20–25% at over 700 sites across the state and thereby save about $43 million annually in energy costs. The commonwealth sold $100 million worth of green municipal bonds to eight to ten institutional investors. They could have sold more, as the sale was oversubscribed by about 30%.[177]

The global green bonds market grew to $160 billion (as of 2017) with three nations—the U.S., China, and France—issuing more than half of the bonds. Some examples of green bonds just in California were for mass transit, water, sewer, and renewable energy projects. The Los Angeles Metropolitan Transportation Authority issued a $471.3 million green bond in October 2017 to underwrite a variety of mass transit projects. The City of Los Angeles and the California Infrastructure Economic Development Bank both issued bonds totaling more than $400 million to fund improvements to water and sewer systems. San Francisco's Bay Area Rapid Transit System issued $384.7 million in green bonds to fund public transportation network improvements, while the San Francisco Public Utilities Commission has issued multiple series of green bonds for clean water and renewable energy projects.[178]

Green bonds provide a growth area for municipal and state governments looking to fund a pipeline of sustainability projects. Many cities maintain a wish list of projects in renewable energy, energy efficiency, sustainable waste management, sustainable land use, biodiversity conservation, clean transportation, clean water, and various climate adaptation projects. Green bonds offer an opportunity for investors' growing desire to diversify their portfolios and invest in sustainability.

Steve Grossman, former Treasurer of Massachusetts, said that these green bonds help investors looking to develop "part of their portfolios for projects that are sustainable or have a green or environmentally sound mandate."[179]

NON-PROFIT SECTOR FUNDING

Potential funding for non-profit sector sustainability projects involves more than soliciting donations and private foundation grants. Other options include low-interest loans from Community Development Financial Institutions, externality fees, self-funding, a fee-for-service structure, and government grants, among others. Here are some examples of where various non-profits have found funding to support their operations.

Zero Net Energy Retrofits—Single-Family Homes

Funding need: capital to make new loans

Enhabit, a 501(c)3 non-profit, identifies energy efficiency opportunities in single-family homes in Portland, Oregon, and connects them with contractors and financing. Some homeowners repay the loans through their Portland General Electric, Northwest Natural Gas, or Pacific Power utility bills, a process called on-bill repayment. Project costs vary from minor fixes to comprehensive energy efficiency retrofits that average about $15,000. In most cases, homeowners save enough money on energy to offset all or part of their payments.

About half of Enhabit customers finance their projects through one of Enhabit's partner lenders. One of those partner lenders, Craft3, sold a portfolio of energy efficiency loans to Self-Help Credit Union so that Craft3 had more cash to recycle into new loans. Through two loan sales in 2013 and 2015, Self-Help purchased 1,716 loans originating from Craft3 and totaling $22 million.[180]

Transportation Management Associations
Funding need: free transit passes for low-income workers and consulting labor

Seventy percent of service workers who commute into downtown Palo Alto, California, drive single-occupancy vehicles. Many of these employees work in restaurants, shops, and hotels in minimum-wage jobs and already own cars. Instead of asking them to buy a bus or train pass on top of the expenses they already incur to keep a personal vehicle running, the Palo Alto Transportation Management Association decided to give away free transit passes to downtown service sector workers. These low-wage service workers then save money on variable costs associated with driving: gas, maintenance, and repairs.

Funding for the free transit passes comes from a combination of downtown parking garage fees and donations from large private employers in the area. The use of these parking fees to pay for transit passes is like using sticks to fund carrots: parking fees emit a pricing signal that discourages people from driving into downtown and uses that funding to help low-wage service workers make a switch to low-carbon mobility options.

Fees like these are sometimes referred to as "feebates" (the Rocky Mountain Institute credits energy efficiency advocate Art Rosenfeld as the one who coined the term). The basic idea as it applies to vehicle purchases is simple: levy a surcharge on people who buy fuel-inefficient vehicles (the "fee"), then reward buyers of fuel-efficient vehicles with a rebate (the "bate"). By sending a price signal to buyers up front, feebates speed up the production and adoption of more efficient vehicles, save people money, reduce oil consumption, and reduce greenhouse gas emissions.[181]

Artistic Upcycling and Salvage
Funding need: labor

The non-profit St. Vincent de Paul hired a full-time fashion designer with the hope that she would be able to cover the cost of her labor with additional store revenue from her designs. Mitra Chester's eye for donations that would sell, along with her innovative designs, raised store revenues from $500/day to $1,500/day. The thrift store's decision to self-fund the hiring of a fashion designer more than paid for itself.

Reverse Catering
Funding need: labor and operating costs

RePlate is able to hire drivers because they charge for pick-ups. With this fee-for-service revenue stream, they are able to pay drivers $20 per hour. Charges of $40 per pick-up cover labor costs for drivers and office rent but do not cover many of their operational costs. As a 501(c)3 non-profit, RePlate receives grants from private foundations to cover expenses such as website maintenance, cloud storage, and invoicing and payroll platforms, all of which have allowed the non-profit to scale.

Community Kitchen
Funding need: labor

As a non-profit that looks for innovative ways to reduce food waste and divert surplus food in the Bay Area, Food Shift primarily relies on donations and grants for their labor and operational costs. Revenue streams from public speaking engagements and hosting events also provide income, but they are excited about growing other revenue sources such as food service contracts for the Alameda Point Collaborative project.

In 2017, about 8% of Food Shift's revenue came from food service contracts and catering gigs. They plan to increase this number as they move toward becoming financially self-supporting.

Business Services Supporting Small Organic Farms
Funding need: consulting labor

Kitchen Table Advisors provides business planning and financial management consulting to small farms and ranches free of charge. Financial support from individuals, businesses and small family foundations pays for the organization's work to grow the number of financially sustainable small, organic farms, and ranches in Northern California.

Carbon Farming
Funding need: labor and supplies

The Marin Carbon Project (MCP) is a consortium of agricultural agencies, non-profits, and producers working to rebuild topsoil and sequester carbon through various carbon farming techniques. Members of the MCP have received government grants from various state and federal agencies. California's Healthy Soils Initiative provided a partial incentive for compost purchased from state-certified facilities on an annual grant basis and for other carbon farming practices. MCP has also secured funding from the U.S. Department of Agriculture's Natural Resources Conservation Service, which includes the Environmental Quality Incentives Program, to help cover amendments for farm nutrient management. In the Frequently Asked Questions section of their website, the MCP also suggests that other non-profits, farmers, and ranchers looking for funding for carbon farming develop carbon farm plans so they can take advantage of funding opportunities through their local Resource Conservation District or land trust.

ADDITIONAL FUNDS NEEDED

With such a large kit of financial tools to choose from, options to fund green projects abound. However, I'd like to see two particular funds developed specifically for social enterprises that provide professional services: a Great Pivot Crowdfund similar to SheEO's model and a Great Pivot Investment Fund that invests in a wide variety of private sector projects yielding an attractive average return.

SheEO is a crowdfunding community that organizes investors into "cohorts" of 500 women "activators." To start, each investor contributes $1,100 to her cohort as an Act of Radical Generosity. Then, every year, each cohort selects five women-led businesses and offers loans at 0%

interest over five years. All ventures that receive investment funding are revenue generating and their missions are to create a better world, either through their business models or the products and services they offer. As the loans are paid back over five years, the funds are loaned out again, thereby creating perpetual financing that benefits future generations. Since 2015, SheEO has invested $2 million in ventures created by 32 female entrepreneurs in the U.S., Canada, and New Zealand. The goal of the organization is to grow the fund to $1 billion, with one million investors supporting 10,000 women-led ventures.

One example of a business that received SheEO funding is Loliware, a startup whose straws resemble colorful plastic but are actually made from seaweed. The founders of Loliware, Chelsea Briganti and Leigh Ann Tucker, have their eyes on replacing a large portion of the 500 million single-use, disposable, plastic straws used daily in the U.S. with their easily biodegradable and edible versions. Funding for Loliware has also come from the Closed Loop Fund, Shark Tank judge Mark Cuban, and a Kickstarter campaign, among others.

In SheEO's spirit of radical generosity, we can also establish a Great Pivot Crowdfund dedicated to activating private sector sustainability projects. The crowdfund should be organized on a regional basis to accelerate development of local economies. This concessionary investing tool would charge no interest or low interest to ensure projects with strong social and environmental benefits receive the startup funding they need.

For projects offering a competitive or above-competitive rate of return, we can also create a Great Pivot Investment Fund to accelerate development of businesses offering professional services. Venture capital and private equity already actively seek out and fund software and products that can scale quickly. Fewer financial tools are available for businesses offering professional services that will build a sustainable future but which will not necessarily deliver spectacular returns. The Great Pivot Investment Fund would bundle projects with various returns to deliver a competitive return on investment.

Entrepreneurs need funding. Professional investors and non-accredited investors want projects they can invest in to build the future they already envision. Many different financial tools are available to build a sustainable future, and we are limited only by our imagination about how to use the tools currently available.

SCALING UP EXISTING PROGRAMS

For now, merely scaling up the funding tools we have and replicating them across the nation would help expand the number of meaningful jobs. In 2015, Judy Wicks, the founder of White Dog Café in Philadelphia and cofounder of the nationwide Business Alliance for Local Living Economies, decided to found a group called Circle of Aunts and Uncles, whose mission is to provide micro-loans and social capital to entrepreneurs without "friends and family" resources. As an entrepreneur herself, Wicks knows that money and advice are not the only things entrepreneurs require. "They need a supportive community who will buy from them and tell their friends," she said. "The aunts and uncles are creating an energy field around local purchasing and the rebuilding of our regional economy."

When Wicks started Circle of Aunts and Uncles, she recruited 35 of her friends to provide loans, contacts, and advice to young Philadelphia entrepreneurs who otherwise would not have access to the financial and social capital resources they need to grow their businesses. Each of the "Aunts and Uncles" funders made initial contributions of $2,000 to help build the fund, then make subsequent donations of $1,000 annually.

To date, the group has given out more than $100,000 to twelve different local businesses, most owned by women and/or people of color. The three-year loans, up to $12,000 each, are paid back at 3% interest.

Wicks explained, "When we make loans to our entrepreneurs, we are also helping to build our local food system and economy. This builds local self-reliance on basic needs, one of the most important things we can do to both mitigate and prepare for climate change."

At the end of the day, this work is about more than survival. Circle of Aunts and Uncles is helping create the types of local businesses that once formed the foundation of community life. "Knowing who bakes our bread, makes our clothes, and brews our beer brings happiness," Wicks said. "The aunts and uncles and 'nieces and nephews' are co-creating our local economy and the community we all want to live in."

CHAPTER HIGHLIGHTS

- Financial tools for private sector projects include boot-strapping, crowdfunding, lines of credit from community development loan funds, direct public offerings, and private equity.
- To fund sustainability projects, the public sector has used financial tools such as carbon impact fees, landfill fees, feebates, utility taxes, parcel taxes, and green bonds.
- Financial tools the non-profit sector uses for projects include self-funding, fee-for-service, parking fees, bundling and selling of energy efficiency loans to free up capital for new loans, crowdfunding, philanthropy, and government grants.

11

DELIVERING ON THE PROMISE

OF MEANINGFUL WORK

*"Good management is the art of making problems
so interesting and their solutions so constructive that
everyone wants to get to work and deal with them."*
—Paul Hawken

One night in October 2017, as a fast-growing wildfire burned near Napa, California, Pete Gavitte and his first mate, paramedic Whitney Lowe, flew continuous helicopter flights to save as many people as they could. An eighteen-year veteran with the California Highway Patrol, Gavitte wore night vision goggles to find people fleeing the fires. At one point, he saw residents throwing their belongings into their cars, jumping in, and heading down Atlas Peak Road, the only way out of the rural community.

"We could see this huge, three-mile-and-growing line of fire coming toward all these folks that probably couldn't see it," Gavitte said. He shined his searchlight on drivers to get their attention and prompt them to stop. Over the next seven hours, he and another team on a second helicopter flew in winds gusting up to 70 miles per hour, racing to extract people from the burning hills. Lowe said that, in a rush to get everyone out of harm's way, the helicopters dropped people off and departed again in just 45 seconds. Among the later rescues was a group of six vineyard workers who spoke no English.

The helicopters made about twenty trips, which became slower and bumpier as conditions deteriorated. "I'd tell them, 'sit down, sit on the floor, it's going to be bumpy,'" Gavitte said. There was little talk on board. Some took photos of the fire as they soared overhead. "I've never been shaken so much in a helicopter in six years of flying with the highway patrol," said Lowe. By the end of the night, the two helicopter crews saved 42 people, five dogs, and a cat from the wildfire, which over the next few days would kill 21 people and destroy 3,500 homes and businesses.[183]

Wildfires in California, which used to occur primarily in fall at the end of the dry season, now happen year-round. Throughout the country, increasingly severe natural disasters fueled by climate change are negatively impacting state budgets. States are looking to the federal government for financial help, but federal emergency response funding isn't enough: cities and states need to upgrade their infrastructure to be able to bounce back from the shocks imposed by future natural disasters.

Over the past few decades, the destructive power of droughts and storms has increased. In October 1994, some areas of Houston received up to 29 inches of rain during Hurricane Rosa. The storm shut down the city, turning the highways into bathtubs. In an attempt to contain fires from broken pipes at oil refineries and chemical plants along the Houston Ship Channel, emergency responders called on the plants to shut down their pipelines. When one executive claimed they could not, as the plant would lose $1 million per hour, the emergency responder told the executive to turn on the news. "Which channel?" the executive asked. "Any channel," the emergency responder said calmly. All three news stations at the time were running live coverage of the aflame Houston Ship Channel. "Shutting them down, sir," was the response.

At the time, that storm seemed devastating. In August 2017, Hurricane Harvey rainfall in Nederland, Texas, east of Houston, peaked at 60 inches.

FUNDING AND PROGRAM PIVOTS TO BUILD A SUSTAINABLE FUTURE

The main reason I wanted to write this book is that each time I finish teaching my fifteen-hour course, "Managing Sustainable Change in an Organization," at the University of California Berkeley Extension, a few students approach me and ask for advice about how to get into sustainability work. While thinking to myself, *I wish there were more sustainability jobs, so everyone who wants one could have one*, I proceed to tell them about the openings of which I am aware.

I meet so many people who dislike their jobs and want to switch to sustainability work. One time, I was working on my laptop in a café while wearing a Women in Cleantech and Sustainability T-shirt when a stranger approached. She said she worked in tech but didn't like her job. She asked where Women in Cleantech and Sustainability met and how she could find a job in cleantech. Another person I met was working as a full-time contractor at Facebook but really wanted to be doing sustainability work. I needed help with a sustainability research project, so she worked for me part-time for a few months, which later helped her secure a sustainability job. People like these hunger for meaningful work that gives their lives purpose.

To accomplish this important transition and create these meaningful jobs, we first need to allocate funding. The search for such funding starts in some of the most well-funded areas of the public and private sectors. By reallocating just a fraction of this funding, we will ignite the creation of millions of meaningful jobs in the private, public, and non-profit sectors.

Pivot (a) Create Externality Fees to Fund Work

Externality fees can be a key source of funding for meaningful jobs. A seal that becomes entangled in a fishing net is an example of a negative externality. Litter on the side of the road is another example. When fossil fuels are burned, they accumulate in the global atmosphere and cause climate change, which is a burden on victims of hurricanes and wildfires as well as the future generations who will be affected. These are all externalities where the victims did not ask to bear the costs imposed on them.

Externality fees can address these problems by accomplishing three things:

1. cleaning up the problem caused by the negative behavior
2. paying for green jobs to create a more sustainable alternative
3. discouraging negative behavior

Externality fees already exist. Ten states have deposit-refund systems for bottles and cans; these deposits deter littering and provide funds for equipment and systems to ensure bottle and can recycling. Some states and cities levy plastic bag fees. A five-cent, ten-cent, or 25-cent fee on a plastic bag at the grocery store encourages people to avoid taking a plastic bag and bring their own reusable bag. California has a Cap & Trade system that levies fees on large emitters of greenhouse gases; these fees go into a fund used for projects that will result in greenhouse gas reductions and improved environmental health. Table 6 shows what the funds were used for during the 2016–2017 fiscal year.

Expanding the number of externality fees similar to California's Cap & Trade fee would discourage environmentally destructive behavior, fund vital sustainability projects, and help us create meaningful jobs. We can apply the idea of externality fees to many different environmental problems that negatively affect people, wildlife, air quality, water quality, and land. Here are some additional externality fees we could levy and what we could accomplish with the funds:

The twelve externality fees in Table 7 are intended to spur discussion about the many other externality fees governments could levy in order to fund meaningful work. Just floating this idea and discussing it through social media, mass media, and alternative media may encourage those that create negative externalities to evolve. The threat of a new fee on pollution and environmental degradation may give corporate executives the cover they need to make the internal changes in support of sustainability they had been wanting to make all along.

Pivot (b) Harness the Power of the Private Financial Sector

The financial sector also has an important role to play in diverting funding from the dinosaur economy into a new green economy. Like lumbering old dinosaurs, most fossil fuel companies, automotive manufacturers, garbage haulers, and utilities have been slow to evolve, even as people working in these industries push for sustainable change. Fossil fuel companies can become energy companies and invest in renewable energy. Automotive companies can become low-carbon mobility providers. Garbage haulers could become recycling logistics and zero-waste firms. Utilities can provide new models for delivering clean energy.

We contribute to the river of funding flowing toward the dinosaur economy each time we pay our bills or put money in our retirement fund, thereby perpetuating the environmentally destructive behavior of the industries from which we purchase goods and services. We pay our credit card bills, which send funds to oil companies. We also pay car payments, garbage bills, and utility bills, all of which encourage non-renewable energy, extractive practices, and waste generation that many of us would like to stop.

This is like using a dishwasher that doesn't clean well. After the machine finishes running, we unload the dishes to find food particles and spots on our dishes. The dishwasher didn't perform its intended job, but we don't feel like handwashing the dishes or calling a repairperson, so we tolerate a suboptimal result.

Our retirement fund investment choices, limited as they are, also support extractive industries and companies that contribute to the take-make-waste economy. Retirement investment funds for non-professional investors do not offer opportunities to invest in projects such as commercial zero net energy retrofits, solar emergency microgrids, Mobility-as-a-Service apps, or ugly produce distribution. Investments in exciting projects like these are happening at a different level.

A small section of the financial sector called impact investing is doing some compelling work. One example is Farmland LP, a sustainable farmland investment company that buys commodity crop farmland and adds value by securing organic certification, investing in infrastructure, and increasing crop diversity. With a minimum

TABLE 6: CALIFORNIA'S CAP & TRADE SPENDING
THROUGH 2016-2017 (IN MILLIONS $)

PROGRAM	AGENCY	2013-14	2014-15
High-speed rail	High-speed Rail Authority	—	250
Affordable housing/sustainable communities	Strategic Growth Council	—	130
Low-carbon vehicles	Air Resources Board	30	200
Transit and intercity rail capital	Transportation Agency	—	25
Low-income weather-ization and solar	Community Services and Development	—	75
Transit operations	Caltrans	—	25
Transformational climate communities program	Strategic Growth Council	—	—
Agricultural energy and efficiency	Food and Agriculture	10	25
Sustainable forests and urban forestry	Forestry and Fire Protection	—	42
Green infrastructure	Natural Resources Agency	—	—
Waste diversion	CalRecycle	—	25
Water efficiency	Dept. of Water Resources	30	20
Wetlands and water-shed restoration	Fish and Wildlife	—	25
Active transportation	Caltrans	—	—
Black carbon woodsmoke	Air Resources Board	—	—
Other technical assistance and administration	Various	2	10
Totals		**$70**	**$852**

2015–16	2016–17	TOTAL
458	250	958
366	200	696
95	363	688
183	235	443
79	20	174
92	50	167
—	140	140
40	65	140
—	40	82
—	80	80
6	40	71
20	—	70
2	—	27
—	10	10
—	5	5
14	24	50
$1,354	**$1,522**	**$3,800**

Source: California Legislative Analyst's Office[184]

TABLE 7: PROPOSED EXTERNALITY FEES AND
THEIR POTENTIAL USE

NO.	ACTIVITY	COST BEARER
1	Burning fossil fuels	Atmosphere, people with asthma, future generations
2	Driving single- occupancy vehicle	People stuck in traffic
3	Suburban sprawl development	Wildlife, people stuck in traffic
4	Demolition waste	Natural areas
5	Single-use disposables	Marine life in coastal areas
6	Single-use cardboard mailing boxes	Municipalities
7	CAFO methane emissions	Atmosphere, area residents breathing the air
8	CAFO waste lagoons	Waterways, area residents breathing the air
9	Overfishing	Fisheries, food web, future generations
10	Pesticide use	Wildlife
11	Synthetic fertilizers	Waterways
12	Food waste	Atmosphere, future generations

EFFECTS	FEE NAME	USE
Climate change, asthma attacks	Carbon fee	Carbon sequestration, energy efficiency, renewable energy
Time away from family and friends	Congestion pricing	Mass transit upgrades, alternative transportation incentives
Loss of habitat, time away from family and friends	Development fee	Expand wildlife habitat, create more national parks and wildlife sanctuaries, mass transit upgrades, alternative transportation incentives
Degradation of natural areas	Demolition fee	Regional database of salvaged building materials
Wildlife injury and death	Single-use disposables fee	Marketing program encouraging reusables
Deforestation	Single-use mailer fee	Reusable mailer reverse logistics system
Climate change, asthma attacks, nuisance	Methane fee	Carbon sequestration, research into methane emission reduction measures for livestock
Degraded waterway quality after severe storms	Waste lagoon fee	Waterway restoration, wildlife restoration
Collapse of marine food web	Fees on wholesale and retail fish purchases	Enforcement of marine wilderness areas, lost and abandoned fishing net retrieval
Collapse of bottom of food web	Pesticide fee	Integrated pest management education and technical assistance
Eutrophication of waterways, marine wildlife collapse	Synthetic fertilizer fee	Organic farming education and technical assistance
Climate change	Food waste fee	Surplus food waste recovery

investment of $500,000, the fund is geared toward accredited investors and institutional investors—and the impact investing community wants more opportunities like these to invest in.

To accelerate development of the green economy, we need people to step up and create more business plans and people to organize new community investment funds. Then we can put the muscle of accumulated wealth to work. Private foundations control $800 billion. In 2015, $3.5 trillion was sitting on the balance sheets of U.S. nonfinancial companies, according to the Federal Reserve. Hedge funds and private equity cash holdings add another $10 trillion of idle capital.[185] Global investors managing $32 trillion in assets, who signed the "2018 Global Investor Statement to Governments on Climate Change," want to invest in a decarbonized economy. If we encourage more aspiring green entrepreneurs to develop business plans and link them up with people who want to invest in a sustainable future, we will be charging the batteries of a green economy that could then accelerate down a much smoother, more open and sustainable highway.

Pivot (c) Establish Green Entrepreneur Basic Income

When I counsel people who want to move into sustainability work, I ask, "How entrepreneurial are you? There is so much sustainability work that requires people to take risks and make it happen." Invariably, they respond that of course they are entrepreneurial. My next question is, "How much of a financial runway do you have to support yourself while you set up a social enterprise?" That's where potential entrepreneurs get stuck.

Establishing a new business or non-profit takes at least twelve months. Between registrations, setting up a bank account and accounting system, securing funding, building a website, doing marketing, and other tasks, starting a business or non-profit can be a full-time job.

More people would be comfortable with the prospect of setting up a new business if they had a way to pay for basic expenses.

At a time when many thought leaders talk about universal basic income to help people meet their basic needs for an indefinite amount of time, green entrepreneur income would provide basic income for a limited amount of time to support the development of new small businesses and non-profits. Doing so would give people brave enough to attempt something new sufficient time for trial and error to ensure their fledgling organization successfully takes flight.

Pivot (d) Develop Green Entrepreneur Accelerators

As a society, we value entrepreneurs, but launching a social enterprise can be hard, lonely work. By creating regionally focused communities of green entrepreneurs, we will provide the support and access they so badly need to accelerate to profitability. Simply creating a network and providing access to other people going through similar struggles is a strong start.

Accelerators already boost startups working in ecosystems such as cleantech software and scalable digital products, offerings that promise quick growth. One such accelerator, Clean Tech Open, brings together dozens of clean tech entrepreneurs for a weekend of skill- and community- building, after which the group meets on a regular basis to make progress on customer discovery, to fine-tune their elevator pitches and develop business plans, and ultimately to pitch to venture capitalists for startup funding. As well, Small Business Development Centers offer online and in-person classes on business structure, accounting, business plans, marketing, and sales, among other areas, but for green professional-services entrepreneurs, few communities exist where visionaries can come together and develop business skills.

I would like to see a hybrid accelerator offer SBDC-type business classes held on a weekend specifically for entrepreneurs launching green, professional service businesses. Starting up the community over one intense weekend and then meeting regularly with other members of the community (either weekly or monthly) would speed up the implementation time-frame and edge the ventures closer to launch.

Social entrepreneurs are an untapped economic force. The Skoll Foundation, which supports social entrepreneurs, sees the potential to "forge a new, stable equilibrium that unleashes new value for society, releases trapped potential, or alleviates suffering."[186] If adequately resourced and supported, social entrepreneurship has the power to drive large-scale change.

Pivot (e) Expand Green Job Training

For people who want to do meaningful work but not necessarily be the entrepreneur launching the business, training to learn specific new job skills will help set them up for success. We also must make sure jobs will be available when people finish job training. Dr. Dan Kammen, founding director of the UC Berkeley's Renewable and Appropriate Energy Laboratory and climate adviser for the Obama and Trump Administrations, cautions that "a job training program that's not linked up to specific industries with documented demand for labor never works. But if you have a federal mandate for clean energy, and you have job training in association with industry, there is a big success route."

To ensure demand exists for the skills the job training provides, government helps in a few corollary roles. When government agencies set goals, develop ordinances, and provide grant funding, they provide the certainty the private sector needs in order to invest as well as instill confidence in entrepreneurs that there will be a market for the goods or services they will offer. The City of Portland's joint deconstruction ordinance, grant program, and deconstruction training provide an example of how synergies between government and industry results in green job creation.

Pivot (f) Create Portable Benefits

Portable benefits like health insurance also supports aspiring entrepreneurs. Many people stay in a job they don't like just so someone in the family has health insurance, and people working for large employers have most of their health insurance costs covered. For those considering starting a social enterprise, the prospect of paying the full cost of health insurance keeps people shackled to jobs they would rather leave. Being able to continue benefits like health insurance even after a change in employment will give people the courage to strike out on their own.

In 2017, Senator Mark Warner (D-VA) introduced Senate Bill 1251, Portable Benefits for Independent Workers Pilot Program Act, and Representative Suzan DelBene introduced related House Bill 2685 to test and evaluate innovative portable benefit designs. Both understand that having benefits allows people in the working world to take risks.

Taken together, these six funding and program pivots will help break through many of the barriers that hold back potential entrepreneurs. As a society, we can do more to support potential entrepreneurs with temporary basic income, portable benefits, stronger communities and networks, and investment funding to prove that our country values those with entrepreneurial spirit.

PLANNING FOR LIFE BEYOND THE INTERREGNUM

The time we are living in is truly special. With recent political, economic, social, and environmental upheavals, it feels like this period is an *interregnum*, meaning a time between kings. Originally, the term referred to a time when a throne was vacant and chaos reigned as warlords battled to fill the power vacuum. Our time feels similar in that something old is dying away and something new is waiting to be born. We once had a stable climate, a secure middle class, and a functioning federal government. All of that has been upended, and unpredictability feels like the new normal.

To stabilize the climate, rebuild the middle class, and redefine government's role to serve the needs of a majority of its citizens, we are called on to be the architects and builders of a sustainable future. The private, public, and non-profit sectors all play vital roles as we shift financial and personnel resources to where they are needed, and in the process of doing so, we will deliver on the promise of creating meaningful work.

CHAPTER HIGHLIGHTS

- Negative externalities can be fixed by levying externality fees to discourage the negative behavior, clean up the problem, and promote sustainable methods.
- A river of funding flows toward dinosaur industries in the form of monthly bills and retirement investments. Diverting a small stream of this funding to finance sustainability projects will give these projects the capital they need to scale up.

- Program pivots to develop meaningful jobs include development of green entrepreneur accelerators, green entrepreneur basic income, job training, and portable benefits.

12

THE GREEN NEW DEAL

"The core of [the Green New Deal] is leave no person behind. We cannot move forward into our future unless every single American is considered part of that."
—Alexandria Ocasio-Cortez, Representative from New York's 14th Congressional District

In September 2017, two days after Category 5 Hurricane Maria ripped through Puerto Rico, Major Frank Noe received a call and was asked if he was available to fly emergency relief supplies to San Juan for the Air National Guard. Noe said yes and within 24 hours was flying a C-17 cargo plane full of generators, Humvees, and National Guard staff and volunteers for indefinite emergency response deployment.

Heading south from the U.S. mainland, pilots like Noe are accustomed to having Positive Radar Control, in which they talk to the control tower and receive clearance to land. Once Noe and his crew passed over the Grand Turk island in the Caribbean, the radio fell silent. Due to power outages, no one on the ground was available to help them land the C-17 at the San Juan airport.

Upon arrival at sunset, Noe saw flattened buildings and few lights. That was when he realized what a big impact the hurricane had had on the island. He dropped his cargo and immediately flew back to Florida.

The next day he made a second flight to San Juan, this time with a 100,000-pound cargo of food and water on pallets. When asked what that experience felt like, he said, "It's great to bring equipment and people that are needed for logistical support, but bringing food and water is even better because it's going directly to the people. You're bringing them life." Noe described the work of delivering initial relief as "an amazing feeling."

Wouldn't it be great if all of us felt similarly about our jobs, that we are being paid for vitally important work that makes a difference? As mentioned in the first chapter, too many people feel trapped in a "bullshit job," one so unnecessary and unfulfilling that even the employee cannot justify it. Now consider the converse: employee engagement in the world of work would be much higher if the majority of people enjoyed a form of paid employment that confers a sense of purpose, happiness, and satisfaction that enriches the life of the employee as she or he works to improve a social or environmental condition.

Imagine what the work world would be like if more people lived in a state of *ikigai*, working at the intersection of doing a job that utilizes their strongest talents and abilities and which the world needs and someone will pay them to do.

As a society, we not only have the opportunity to develop meaningful jobs—we have an obligation to proactively plan for them. In fact, given the seismic shifts underway in the world of work, we don't have a choice: multiple disruptions are contributing to an ongoing crisis, from the recent disruption of outsourcing to the current disruption of the gig economy to the impending disruption posed by increased automation.

In the transportation sector alone, a portion of the 4.2 million people who drive for a living face the prospect of potentially losing their jobs to autonomous vehicles in the next decade. One of the visionaries and implementers of the future of autonomous vehicles is Tesla CEO Elon Musk, who knows that AVs will lead to mass unemployment and thus suggests that we establish universal basic income to maintain social and economic stability, saying:

> " . . . I think maybe these things do play into each other a little
> bit, but what to do about mass unemployment? This is going

to be a massive social challenge. And I think, ultimately, we will have to have some kind of universal basic income. I don't think we're going to have a choice."[187]

Automation will yield dividends that we could use to fund universal basic income payments, or we could use that money to create green jobs. Let's give people a choice. In building a decarbonized, circular economy and restoring nature, plenty of jobs will be created. Let's be proactive about the impending future of automation not resigned. Inventor and visionary Buckminster Fuller urged us to be "the architects of the future, not its victims."

BUILDING A BRIDGE TO A DECARBONIZED FUTURE

Coal miners in Appalachia are the poster children for people stuck between old and new: automation has shrunk their ranks, and our economy is decarbonizing. Consider that in 1923 there were 863,000 coal miners; in 2018 there were 52,900.[189] Between 2011 and 2017, Appalachia lost 33,500 mining jobs alone. The coal industry continues to try, through automation, to stay profitable for a little while longer as the economy switches to cleaner forms of energy.

Coal miners left are trying to hold on—and understandably, because the industry actually pays quite well. Someone with a high school diploma working six or seven days a week and logging overtime hours can make $100,000 to $140,000 per year. In rural Pennsylvania, that's enough income to buy a nice house, four-wheel-drive truck, motorcycle, vacations, and to put the kids through college and support a spouse. Some miners don't even venture down the mine shaft and instead work aboveground, directing machines with control panels and joysticks.

Greene County, Pennsylvania, a county with 36,000 people in the southwest corner of the state, hasn't felt the blow that some other coal-producing areas have, such as West Virginia and Kentucky, because of the uptick in foreign demand for metallurgical coal used in producing steel. Employment in the coal industry runs in bursts though. The industry mines a particular coal seam until all the coal has been extracted, then transfers or lays off miners. Greene County Commissioner Blair Zimmerman explained, "We just lost 250 jobs when the Cumberland, a deep longwall mine, ran out of coal." However, given the cycles of employment and unemployment, "Coal miners are an optimistic bunch. They think they're going to get called back," Zimmerman said.

The federal government has been trying to bolster against this instability. Between 2015 and 2017, the U.S. Department of Labor gave the Pennsylvania Department of Labor $2 million to train coal miners for new skills. Sign-up rates for the training, however, were less than 20%. The Appalachian Regional Commission, a federal and state partnership whose aim is to strengthen the region's economy, is backed with $1.4 million for teaching computer coding to laid-off miners in Greene County and in West Virginia. They have only signed up 20 people for 95 slots. Not a single worker has enrolled in another program launched in 2017 to prepare ex-miners to work in the natural gas sector.

A *Reuters* article entitled "Awaiting Trump's coal comeback, miners reject retraining" conducted more than a dozen interviews with former and prospective coal miners, career counselors, and economic development officials and discovered four main reasons for the reluctance to go through new skills training. People interviewed explained that:

1. mining pays well,
2. they are unfamiliar with other industries,
3. there's no income during training, and
4. there's no guarantee of a job afterward.[190]

Resistance to retraining is understandable but leaves economic development managers concerned about the future of their local workforce. Zimmerman would like to be able to attract a large manufacturing plant to his county, given the skills miners possess for working with electrical equipment, hydraulics, trucks, and heavy equipment.

I shared the list of 30 meaningful work projects in *The Great Pivot* with Zimmerman as food for thought. As the coal industry automates further, would some of the miners be interested in retrofitting buildings for zero net energy, doing building deconstruction, making furniture from salvage timber, working in manufacturing that uses recycled content, or restoring wildlife habitat? (It was a bit surreal talking to this 40-year veteran of the coal mines about the need for jobs that sequester carbon and restore nature.) Zimmerman had to mull this over, but the question persists: Can we find money in our economy to pay people green jobs wages competitive with what miners currently make? This is the time to map out a clear path to prosperity for some of the nation's most at-risk areas.

As we plan for the future world of work, the good news about green jobs is that many of them can't be outsourced or automated. Building deconstruction, solar installation, carbon farming, and wildlife restoration must be done by people here in the U.S. Now we just need to figure out which funding and program pivots we want to make to deliver these stable, middle class Great Pivot jobs.

One vehicle to bring about the Great Pivot may be the Green New Deal. As I wrap up the writing of *The Great Pivot*, news articles and fresh blog posts appear daily about this visionary program. It just may be an idea whose time has come, as a poll from the Yale Program on Climate Change Communication found: a survey of 966 registered voters between November 28 and December 11, 2018 reported that 81% of Americans either strongly support or somewhat support a Green New Deal.[191]

In November 2018, the youth-led Sunrise Movement occupied the offices of House Minority Leader Nancy Pelosi, New Jersey Democrat Frank Pallone, and others in Washington, D.C., demanding meaningful action on climate change. Alexandria Ocasio-Cortez, the freshman representative from New York's 14th district, showed up at the sit-in to express support for the live-streamed protest and effectively assumed the mantle of Green New Deal leadership. Since then, dozens of elected officials from Congress have publicly backed the concept.

We are just starting to see the broad outlines of a more muscular and strategic green economy that will be deployed at a scale to match the severity of the potential threat climate change poses. The initial legislative wish list for a Green New Deal includes dramatically expanding renewable energy, upgrading buildings for energy efficiency, and eliminating greenhouse gas emissions from the transportation sector. Creating green job banks where people who want a green job can pick from a list of local vacancies is another idea being proposed, as well as job training, a family-sustaining wage, and providing all people with high-quality, affordable healthcare. This resolution is not a legislative roadmap, it's a non-binding resolution. Even though we don't yet have the details nailed down, the framework appears to have captured the public's imagination. We as a society now have the opportunity to talk through the details so the Green New Deal can become what we need it to be.

New York Times columnist Thomas Friedman, who has been talking publicly about a Green New Deal since he wrote a widely read article about it in 2007, updated the idea in a January 2019 post, proposing that the Green New Deal include the "four zeros," a term energy innovator Hal Harvey outlined as including:

1. **Zero net energy buildings**—buildings that generate as much energy as they use over the course of a year
2. **Zero waste manufacturing**—designing and building products that use fewer raw materials and that are easily disassembled and recycled
3. **Zero carbon grid**—combining utility-scale renewable power generation with consumers installing their own renewable energy, integrated with the electric grid using large-scale storage batteries
4. **Zero emissions transportation**—combining electric vehicles and electric public transportation with a zero carbon grid[192]

Add to this the Great Pivot's ideas of building a circular economy, reducing food waste, and restoring nature, and the Green New Deal could truly transform our economy by creating millions of meaningful jobs.

By focusing on job creation, the Green New Deal brilliantly addresses the twin excuses that reactionaries trot out to delay action on climate change: it will hurt the economy, and we can't afford it. Investing in the transformation of our economy from fossil fuels to renewables, from extractive industries to restorative activities, from a take-make-waste economy to a circular flow of materials, and from goods to services will result in job losses in the old economy. At the same time, we will be creating new jobs in the green economy. The investment that creates these new jobs will become income for families who will spend it in their local area where it multiplies to become income for others. Simultaneously, as we create middle class jobs that cannot be automated or outsourced, we will also create more resilient infrastructure and restore the natural world so that both can withstand upcoming climatic changes.

OUR LEGACY

After World War II, our parents and grandparents taxed themselves at a high rate to invest in roads, bridges, train systems, airports, water treatment plants, and power plants. This infrastructure formed the foundation of our developed nation. It's time to update our infrastructure for this century.

With this investment, we will give birth to the next great industries that will yield deeply satisfying livelihoods. As we commit to this important work, we will be filled with a sense of purpose, connection, and love, because we will know that this will be our legacy. When future generations look back at our time, they will be grateful that we invested in the Great Pivot. In doing so, we will know that our lives made a difference to those in the future, to society, and to those around us, which is all we ever really wanted anyway.

CHAPTER HIGHLIGHTS

- The coal mining industry has shrunk from peak employment of 863,000 workers in 1923 to 52,900 workers in 2018. As automation replaces workers and our economy decarbonizes further, can we create well-paying green jobs for coal miners as an alternative?
- A discussion is underway in our society about the contours of a Green New Deal. The concept is still inchoate, but details will continue to emerge as we discuss it.
- The Green New Deal addresses the twin excuses reactionaries raise to delay action on climate change: it will hurt the economy, and we can't afford it. Creating jobs provides income to people, helps our economy use resources more efficiently, and makes our infrastructure and nature more resilient.

ACRONYM	FULL TERM
AEC	Advanced Energy Communities
AI	Artificial Intelligence
AV	Autonomous Vehicle
BAU	Business As Usual
BLS	U.S. Department of Labor's Bureau of Labor Statistics
BMRA	Building Materials Reuse Association
BP	British Petroleum
BRT	Bus Rapid Transit
C&D	Construction and Demolition
CAFO	Confined Animal Feeding Operation
CCF	100 cubic feet
CDFI	Community Development Financial Institution
CDLF	Community Development Loan Fund
CEF	Central Energy Facility
CEQA	California Environmental Quality Act
DCFC	Direct Current Fast Charging
DPO	Direct Public Offering
EIR	Environmental Impact Report
EV	Electric Vehicle
EVCI	Electric Vehicle Charging Infrastructure
FEMA	Federal Emergency Management Agency
GDP	Gross Domestic Product
HVAC	Heating Ventilation and Air Conditioning
IoT	Internet of Things
IPCC	Intergovernmental Panel on Climate Change
IRS	Internal Revenue Service
ISS	International Space Station
MAAS	Mobility-as-a-Service
MCP	Marin Carbon Project
MFD	Multi-Family Dwellings
NEPA	National Environmental Policy Act
NILF	Not in the Labor Force
OSB	Oriented Strand Board
PAYS	Pay As You Save
PACE	Property Assessed Clean Energy
SBDC	Small Business Development Center
SMB	Small and Medium-sized Businesses
SRP	Stanford Research Park
SV2	Silicon Valley Social Venture
TMA	Transportation Management Associations
USFS	U.S. Forest Service
WWF	World Wildlife Fund
ZNE	Zero Net Energy

BIBLIOGRAPHY

Benyus, Janine M., *Biomimicry: Innovation Inspired by Nature*. New York: HarperCollins, 1997.

Christensen, Clayton M., *The Clayton M. Christensen Reader*. Boston: Harvard Business School Publishing Corporation, 2016.

Daily, Gretchen C., *Nature's Services: Societal Dependence on Natural Ecosystems*. Washington, D.C.: Island Press, 1997.

Eisenstein, Charles, *Sacred Economics: Money, Gift, and Society in the Age of Transition*. Evolver Editions, 2011.

Eisler, Riane, *The Real Wealth of Nations*. San Francisco: Berrett-Koehler Publishers, Inc., 2008.

Graeber, David, *Bullshit Jobs: A Theory*. New York: Simon & Schuster, 2018.

Hawken, Paul, *Blessed Unrest: How the largest Movement in the World Came into Being and Why No One Saw It Coming*. New York: Penguin Group, 2007.

Hawken, Paul, *Drawdown: The Most Comprehensive Plan Ever Proposed to Reverse Global Warming.* New York: Penguin Books, 2017.

Heath, Chip and Dan Heath, *Switch: How to Change Things When Change is Hard.* New York: Broadway Books, 2010.

Hewitt, Carol Peppe, *Financing Our Foodshed: Growing Local Food with Slow Money.* Gabriola Island, British Columbia: New Society Publishers, 2013.

Hollender, Jeffrey, *What Matters Most: How a Small Group of Pioneers Is Teaching Social Responsibility to Big Business, and Why Big Business Is Listening.* Basic Books, 2004.

Jones, Van, *The Green Collar Economy: How One Solution Can Fix Our Two Biggest Problems.* New York: Harper Collins, 2008.

Kassan, Jenny, *Raising Capital on Your Own Terms: How to Fund Your Business Without Selling Your Soul.* Oakland: Berrett-Kohler Publishers. 2017.

Kimmerer, Robin Wall, *Braiding Sweetgrass: Indigenous Wisdom, Scientific Knowledge, and the Teachings of Plants.* Canada: Milkweed Editions, 2013.

Korten, David C., *The Great Turning: From Empire to Earth Community.* Bloomfield, CT: Kumarian Press, Inc. and San Francisco: Berrett-Koehler Publishers, Inc., 2006.

Kuttner, Robert. *Everything for Sale: The Virtues and Limited of Markets.* Chicago: University of Chicago Press, 1999.

McKenzie-Mohr, Doug, Nancy R. Lee, P. Wesley Schultz, and Philip Kotler, *Social Marketing to Protect the Environment: What Works.* Thousand Oaks, CA: SAGE Publications, Inc. 2012.

Meadows, Donella, *Thinking in Systems: A Primer.* White River Junction, Vermont: Chelsea Green Publishing, 2008.

Mykelby, Mark, Patrick Doherty, and Joel Makower. *The New Grand Strategy: Restoring America's Prosperity, Security, and Sustainability in the 21ˢᵗ Century.* New York: St. Martin's Press, 2016.

Nordhaus, Ted and Michael Shellenberger, *Break Through: Why We Can't Leave Saving the Planet to Environmentalists.* New York: Houghton Mifflin Harcourt Publishing Company, 2007.

Orsi, Janelle and Emily Doskow, *The Sharing Solution: How to Save Money, Simplify Your Life & Build Community,* Berkeley, California: Nolo, 2009.

Raworth, Kate. *Doughnut Economics: Seven Ways to Think Like a 21ˢᵗ Century Economist.* White River Junction, Vermont: Chelsea Green Publishing, 2017.

Rogers, Everett M., *Diffusion of Innovations.* New York: Free Press, 2003 (fifth edition).

Sachs, Jeffrey D., *The End of Poverty: Economic Possibilities for Our Time.* New York: Penguin Group, 2005.

Schumacher, E.F., *Small Is Beautiful: Economics as If People Mattered.* Vancouver, BC: Hartley & Marks Publishers Inc., 1999.

Senge, Peter, C. Otto Scharmer, Joseph Jaworski, and Betty Sue Flowers, *Presence: An Exploration of Profound Change in People, Organizations, and Society.* New York: Doubleday, 2004.

Tasch, Woody, *Inquiries into the Nature of Slow Money.* White River Junction, Vermont: Chelsea Green Publishing, 2008.

Toensmeier, Eric, *The Carbon Farming Solution: A Global Toolkit of Perennial Crops and Regenerative Agriculture Practices for Climate Change Mitigation and Food Security.* White River Junction, Vermont: Chelsea Green Publishing, 2016.

Weisbord, Marvin and Sandra Janoff, *Future Search: An Action Guide to Finding Common Ground in Organizations & Communities.* San Francisco: Berrett-Koehler Publishers, Inc., 2000.

Willard, Bob, *The New Sustainability Advantage: Seven Business Case Benefits of a Triple Bottom Line.* Gabriola Island, BC, Canada: New Society Publishers, 2012.

ENDNOTES

1. Hiroko Tabuchi, "Coal Mining Jobs Trump Would Bring Back No Longer Exist," *New York Times*, March 29, 2017, https://www.nytimes.com/2017/03/29/business/coal-jobs-trump-appalachia.html.

2. Bureau of Labor Statistics, "Databases, Tables & Calculators by Subject," https://data.bls.gov/timeseries/LNS14000000.

3. Christopher Rugaber, "U.S. employers post record number of open jobs in August," *AP News*, October 16, 2018, https://apnews.com/a89dd1a896ba40498fa220dc5f11650e.

4. George M. Reynolds and Amanda Shendruk, "Demographics of the U.S. Military," Council on Foreign Relations, April 24, 2018, https://www.cfr.org/article/demographics-us-military.

5. Peter Wagner and Wendy Sawyer, "Mass Incarceration: The Whole Pie 2018." Prison Policy Initiative. March 14, 2018, https://www.prisonpolicy.org/reports/pie2018.html.

6. Bureau of Labor Statistics, "Table A-15. Alternative measures of labor underutilization," accessed January 11, 2019, https://www.bls.gov/news.release/empsit.t15.htm.

7. Heather Long, "U.S. Has Lost 5 Million Jobs to Outsourcing Since 2000," CNN, March 29, 2016, https://money.cnn.com/2016/03/29/news/economy/US-manufacturing-jobs/index.html.

8. James Manyika, Susan Lund, Michael Chui, Jacques Bughin, Jonathan Woetzel, Parul Batra, Ryan Ko, and Saurabh Sanghvi, "Jobs Lost, Jobs Gained: What the future

of work will mean for jobs, skills, and wages," *McKinsey Quarterly*, November 2017, https://www.mckinsey.com/featured-insights/future-of-organizations-and-work/jobs-lost-jobs-gained-what-the-future-of-work-will-mean-for-jobs-skills-and-wages.

9. Dennis R. Mortensen, "Automation may take our jobs—but it'll restore our humanity," Quartz, August 16, 2017, https://qz.com/1054034/automation-may-take-our-jobs-but-itll-restore-our-humanity/.

10. Wikipedia, "History of the steam engine," accessed January 11, 2019, https://en.wikipedia.org/wiki/History_of_the_steam_engine.

11. History, "Cotton Gin and Eli Whitney," accessed January 11, 2019, https://www.history.com/topics/inventions/cotton-gin-and-eli-whitney.

12. Made How, "Asphalt Paver," accessed January 11, 2019, http://www.madehow.com/Volume-3/Asphalt-Paver.html.

13. Bureau of Labor Statistics, "Contingent and Alternative Employment Arrangements Summary," last updated June 7, 2018, https://www.bls.gov/news.release/conemp.nr0.htm.

14. Freelancer's Union, "Freelancing in America: 2017," accessed January 9, 2019, https://s3.amazonaws.com/fuwt-prod-storage/content/FreelancingInAmerica Report-2017.pdf.

15. Dan Kopf, "Almost all the U.S. jobs created since 2005 are temporary," *Quartz*, December 5, 2016, https://qz.com/851066/almost-all-the-10-million-jobs-created-since-2005-are-temporary/.

16. Elaine Pofeldt, "Are We Ready for a Workforce That Is 50% Freelance?" *Forbes*, October 17, 2017, https://www.forbes.com/sites/elainepofeldt/2017/10/17/are-we-ready-for-a-workforce-that-is-50-freelance.

17. Gallup, "Gallup Daily: U.S. Employee Engagement," accessed January 11, 2019, https://news.gallup.com/poll/180404/gallup-daily-employee-engagement.aspx.

18. Gallup, "Gallup, State of the American Workplace," accessed January 11, 2019, https://news.gallup.com/reports/199961/7.aspx#.

19. Jim Harter and Annamarie Mann, "The Right Culture: Not Just About Employee Satisfaction," Gallup, April 12, 2017, https://www.gallup.com/workplace/236366/right-culture-not-employee-satisfaction.aspx.

20. Cone Communications, "2016 Cone Communications Millennial Employee

Engagement Study," accessed January 9, 2019, http://www.conecomm.com/research-blog/2016-millennial-employee-engagement-study.

21. David Graeber, "On the Phenomenon of Bullshit Jobs: A Work Rant," Strike! Magazine, August 2013, https://strikemag.org/bullshit-jobs.

22. David Graeber, *Bullshit Jobs: A Theory* (New York: Simon & Schuster, 2018), 9-10.

23. David Graeber, "Bullshit jobs and the yoke of managerial feudalism," *The Economist*, June 28, 2018, https://www.economist.com/open-future/2018/06/29/bullshit-jobs-and-the-yoke-of-managerial-feudalism.

24. Michael F. Steger, Bryan J. Dik, and Ryan D. Duffy, "Measuring Meaningful Work: The Work and Meaning Inventory," *Journal of Career Assessment* (00)0 1-16 (2012), http://www.michaelfsteger.com/wp-content/uploads/2012/08/Steger-Dik-Duffy-JCA-in-press.pdf.

25. International Panel on Climate Change, "Global Warming of 1.5°C," accessed January 11, 2018, https://www.ipcc.ch/sr15/.

26. Sarah Sattelmeyer, Sheida Elmi, and Joanna Biernacka-Lievestro, "Typical Family Income Improved in 2016—but Financial Stability Remained Elusive," Pew, October 2, 2017, https://www.pewtrusts.org/en/research-and-analysis/articles/2017/10/02/typical-family-income-improved-in-2016-but-financial-stability-remained-elusive.

27. Angus Deaton, "The U.S. Can No Longer Hide from Its Deep Poverty Problem," *New York Times*, January 24, 2018, https://www.nytimes.com/2018/01/24/opinion/poverty-united-states.html.

28. Peter Wagner and Wendy Sawyer, "Mass Incarceration: The Whole Pie 2018," Prison Policy Initiative, March 14, 2018, https://www.prisonpolicy.org/reports/pie2018.html.

29. Eillie Anzilotti, "Homeboy Recycling Helps Formerly Incarcerated Workers Get On Their Feet With E-Cycling," *Fast Company*, February 13, 2017, https://www.fastcompany.com/3068115/homeboy-recycling-helps-formerly-incarcerated-workers-get-on-their-fe.

30. Amanda Kludt, "Donnel Baird Wants to Build an Actually Ethical Billion Dollar Company," *Eater*, November 9, 2018, https://www.eater.com/2018/11/9/18072638/start-to-sale-donnel-baird-blocpower.

31. Apolitical Group Limited, "Portland creates skilled jobs by banning the demolition of old homes," accessed June 20, 2017, https://apolitical.co/solution_article/portland-creates-skilled-jobs-banning-demolition-old-homes/.

32. Federal Reserve Bank of St. Louis, "Student Loans Owned and Securitized, Outstanding," last updated November 7, 2018, https://fred.stlouisfed.org/series/SLOAS.

33. Yasmeen Abutaleb, "U.S. healthcare spending to climb 5.3 percent in 2018: agency," *Reuters*, February 14, 2018, https://www.reuters.com/article/us-usa-healthcare-spending/us-healthcare-spending-to-climb-53-percent-in-2018-agency-idUSKCN1FY2ZD.

34. Umair Haque, "Soul Care in the Time of Collapse," *Medium*, January 29, 2018, https://umairhaque.com/soul-care-in-a-time-of-collapse-bd3b6e3712b4.

35. Global Footprint Network, "Earth Overshoot Day," accessed January 11, 2019, https://www.overshootday.org/.

36. *The Onion*, "Apartment Broker Recommends Brooklyn Residents Spend No More Than 150% of Income on Rent," September 2, 2016, https://www.theonion.com/apartment-broker-recommends-brooklyn-residents-spend-no-1819579222.

37. S. Pacala and R. Socolow, "Stabilization Wedges: Solving the Climate Problem in the Next 50 Years with Current Technologies," *Science*, Vol. 305, August 13, 2004, 968-972, https://slideplayer.com/slide/6377458/.

38. Stanford University Office of Sustainability, "Stanford University Energy and Climate Plan," Third Edition, rev. September 2015, 37, https://sustainable.stanford.edu/sites/default/files/resource-attachments/E_C_Plan_2015.pdf.

39. U.S. Environmental Protection Agency, "Fact Sheet: Clean Power Plan By The Numbers," accessed January 8, 2019, https://archive.epa.gov/epa/cleanpowerplan/fact-sheet-clean-power-plan-numbers.html.

40. William Driscoll, "Solar could replace 2.73 GW of PacifiCorp coal units, at lower cost," *pv magazine*, August 15, 2018, https://pv-magazine-usa.com/2018/08/15/solar-could-replace-2-73-gw-of-pacificorp-coal-units-at-lower-cost/.

41. Erin Blakemore, "Fifty Years Ago, This Photo Captured the First View of Earth From the Moon." *Smithsonian Magazine*, August 3, 2016, https://www.smithsonianmag.com/smart-news/fifty-years-ago-this-photo-captured-first-view-of-earth-from-the-moon-180960222/.

42. Ashley Strickland, "Scott Kelly from space: Earth's atmosphere 'looks very, very fragile,'" CNN, February 12, 2016, https://www.cnn.com/2016/02/11/health/scott-kelly-space-station-sanjay-gupta-interview/index.html.

43. U.S. Energy Information Administration, "What is U.S. electricity generation

by energy source?" accessed January 11, 2019, https://www.eia.gov/tools/faqs/faq. php?id=427&t=3.

44. The Solar Foundation, "National Solar Jobs Census," accessed January 7, 2019, https://www.thesolarfoundation.org/national/.

45. Conversation with Clark Brockman, Principal of SERA Architects, September 2017.

46. U.S. Census Bureau, "2016 SUSB Annual Data Tables by Establishment Industry," December 2018, https://www.census.gov/data/tables/2016/econ/ susb/2016-susb-annual.html.

47. Kate Murphy, "Parking Spaces That Could Make You Rich," *New York Times*, November 2, 2017, https://www.nytimes.com/2017/11/02/realestate/parking-spaces-that-could-make-you-rich.html.

48. United States Environmental Protection Agency, "Sources of Greenhouse Gas Emissions," accessed January 11, 2019, https://www.epa.gov/ghgemissions/ sources-greenhouse-gas-emissions.

49. Michael Kimmelman, "Paved, but Still Alive," *New York Times*, January 6, 2012, https://www.nytimes.com/2012/01/08/arts/design/taking-parking-lots-seriously-as-public-spaces.html.

50. Donald Shoup, "The High Cost of Minimum Parking Requirements," *Parking: Issues and Policies, Transport and Sustainability*, Volume 5, pp. 87-113, (2014) http:// shoup.bol.ucla.edu/HighCost.pdf.

51. Philip Reid and Nicole Arata, "What Is the Total Cost of Owning a Car?" NerdWallet, October 18, 2018, https://www.nerdwallet.com/blog/loans/ total-cost-owning-car/.

52. Bureau of Labor Statistics, "Table B-3. Average hourly and weekly earnings of all employees on private nonfarm payrolls by industry sector, seasonally adjusted," last modified January 04, 2019, https://www.bls.gov/news.release/empsit.t19.htm.

53. *The Sentinel*, "These 25 Cities Have the Worst Commutes," December 22, 2018, http://hanfordsentinel.com/jobs/these-cities-have-the-worst-commutes-in-america/collection_76fe1af3-5b15-5fd9-bd74-07ad269f4149.html.

54. National Safety Council, "2017 Estimates Show Vehicle Fatalities Topped 40,000 for Second Straight Year," Accessed February 17, 2019, https://www.nsc. org/road-safety/safety-topics/fatality-estimates.

55. Inside EVs, "Bob Lutz Strikes Again: Automakers Should Just Give Up, 'It's All Over'," May 1, 2018, https://insideevs.com/evannex-lutz-autonomous-cars/.

56. Paul Barter, "'Cars are parked 95% of the time'. Let's check!," *Reinventing Parking*, February 22, 2013, https://www.reinventingparking.org/2013/02/cars-are-parked-95-of-time-lets-check.html.

57. Seth Miller, "How Big Oil Will Die," *NewCo Shift*, May 25, 2017, https://shift.newco.co/2017/05/25/this-is-how-big-oil-will-die/.

58. National Highway Transportation Safety Administration, "Vehicle Survivability and Travel Mileage Schedules," DOT HS 809 952 (January 2006), retrieved from https://crashstats.nhtsa.dot.gov/Api/Public/ViewPublication/809952.

59. Fred Lambert, "Tesla battery data shows path to over 500,000 miles on a single pack," *Electrek*, November 1, 2016, https://electrek.co/2016/11/01/tesla-battery-degradation/.

60. The costs and savings associated with the transition to autonomous vehicles involve many variables. Tony Seba's analysis is just one scenario that uses a set of assumptions that may not include variables that are included in other analyses. Also keep in mind that these assumptions focus on urban and suburban areas.

61. Bureau of Labor Statistics, "Heavy and Tractor-trailer Truck Drivers," last updated Friday, April 13, 2018, https://www.bls.gov/ooh/transportation-and-material-moving/heavy-and-tractor-trailer-truck-drivers.htm.

62. Bureau of Labor Statistics, "Delivery Truck Drivers," last updated Monday, June 11, 2018, https://www.bls.gov/ooh/transportation-and-material-moving/delivery-truck-drivers-and-driver-sales-workers.htm..

63. Bureau of Labor Statistics, "Bus Drivers," last updated Friday, April 13, 2018, https://www.bls.gov/ooh/transportation-and-material-moving/bus-drivers.htm.

64. Bureau of Labor Statistics, "Material Moving Machine Operators," last updated Friday, April 13, 2018, https://www.bls.gov/ooh/transportation-and-material-moving/material-moving-machine-operators.htm.

65. Bureau of Labor Statistics, "Taxi Drivers, Ride-Hailing Drivers, and Chauffeurs," last updated Friday, April 13, 2018, https://www.bls.gov/ooh/transportation-and-material-moving/taxi-drivers-and-chauffeurs.htm.

66. National Automobile Dealers Association, "New Report: New-Car Dealership Employment Sets Record in 2016," April 13, 2017, https://www.nada.org/nada-data-2016/.

67. Randal O'Toole, "Rapid Bus: A Low-Cost, High-Capacity Transit System for Major Urban Areas," *Policy Analysis*, Number 752 (July 30, 2014), https://object.cato.org/sites/cato.org/files/pubs/pdf/pa752.pdf.

68. Institute for Transportation and Development Policy, "What is BRT?" accessed January 11, 2018, https://www.itdp.org/library/standards-and-guides/the-bus-rapid-transit-standard/what-is-brt/.

69. Stanford Research Park, "2017 Commute Survey Results," October 17, 2017, https://stanfordresearchpark.com/blog/2017/2017-commute-survey-results.

70. Angie Schmitt, "100 Million Americans Bike Each Year, But Few Make It a Habit," Streetsblog USA, March 4, 2015, https://usa.streetsblog.org/2015/03/04/survey-100-million-americans-bike-each-year-but-few-make-it-a-habit/.

71. Michael Andersen, "No, Protected Bike Lanes Do Not Need to Cost $1 Million Per Mile," PeopleForBikes, May 16, 2017, https://peopleforbikes.org/blog/protected-bike-lanes-do-not-cost-1-million-per-mile/.

72. Lena V. Groeger, "Unsafe at Many Speeds," *ProPublica*, May 25, 2016, https://www.propublica.org/article/unsafe-at-many-speeds. Based on research published by B. C. Tefft, "Impact speed and a pedestrian's risk of severe injury or death," *Accident Analysis & Prevention* 50 (2013), pp. 871-878.

73. Aarian Marshall, "The Stubborn Bike Commuter Gap Between American Cities," September 22, 2018, *Wired*, January 5, 2019, https://www.wired.com/story/cycling-census-commuter-data-gap/.

74. "Electric Vehicle Charging Infrastructure Master Plan," Clean Coalition, April 2018, http://www.clean-coalition.org/site/wp-content/uploads/2018/05/PAEC-Task-6.4-Final-Report-on-EVCI-Master-Plan-27_wb_4-Apr-2018.pdf.

75. City of Palo Alto, "Palo Alto Paves the Way for More Electric Vehicles with Local Renewable Energy," last updated July 20, 2017, https://www.cityofpaloalto.org/news/displaynews.asp?NewsID=4018.

76. National Association of Realtors, "Millennials and the Silent Generation Drive Desire for Walkable Communities, Say Realtors," December 19, 2017, https://www.nar.realtor/newsroom/millennials-and-silent-generation-drive-desire-for-walkable-communities-say-realtors.

77. International Making Cities Livable, "City by City, Block by Block: Building Better Blocks Project," accessed January 12, 2019, http://www.livablecities.org/blog/city-city-block-block-building-better-blocks-project.

78. Michael Wagler, "Inspiring a Community and State to Build a Better Block," Main Street America, April 24, 2018, https://www.mainstreet.org/blogs/national-main-street-center/2018/04/24/inspiring-a-community-and-state-to-build-a-better.

79. Wikiblok, "Dream It. Print It. Build It. Live It," accessed January 12, 2019, http://betterblock.org/wikiblock/.

80. Billy Perrigo, "A Dutch Teenager Had a Dream to Clean Up the World's Oceans. 7 Years On, It's Coming True," *Time*, September 7, 2018, http://time.com/5389782/boyan-slat-plastic-ocean-cleanup/.

81. Hannah Summers, "Great Pacific garbage patch $20m cleanup fails to collect plastic," *The Guardian*, December 20, 2018, https://www.theguardian.com/environment/2018/dec/20/great-pacific-garbage-patch-20m-cleanup-fails-to-collect-plastic.

82. Annie Leonard, "Story of Stuff: Referenced and Annotated Script," accessed January 12, 2019, https://storyofstuff.org/wp-content/uploads/movies/scripts/Story%20of%20Stuff.pdf.

83. Wrap, "WRAP and the circular economy," 2018, http://www.wrap.org.uk/about-us/about/wrap-and-circular-economy.

84. Jerry Powell, "Operation Green Fence is deeply affecting export markets," *Resource Recycling*, April 12, 2013, https://resource-recycling.com/recycling/2013/04/12/operation-green-fence-is-deeply-affecting-export-markets/.

85. Jason Margolis, "Mountains of U.S. recycling pile up as China restricts imports," *PRI*, January 1, 2018, https://www.pri.org/stories/2018-01-01/mountains-US-recycling-pile-china-restricts-imports.

86. Cassandra Profita and Jeff Burns, "Recycling Chaos in U.S. as China Bans 'Foreign Waste'," *NPR*, December 9, 2017, https://www.npr.org/2017/12/09/568797388/recycling-chaos-in-u-s-as-china-bans-foreign-waste.

87. CalRecycle, "2016 California Exports of Recyclable Materials," June 2017, https://www2.calrecycle.ca.gov/Publications/Download/1305.

88. U.S. Environmental Protection Agency, "Sustainable Management of Construction and Demolition Materials," accessed January 12, 2019, https://www.epa.gov/smm/sustainable-management-construction-and-demolition-materials.

89. Megan Cottrell, "Libraries and the Art of Everything Maintenance," *American Libraries Magazine*, September 1, 2017, https://americanlibrariesmagazine.org/2017/09/01/libraries-everything-maintenance-repair-cafe/.

90. Crystle Martin, "Who says libraries are dying? They are evolving into spaces for innovation," *The Conversation*, January 11, 2019, https://theconversation.com/who-says-libraries-are-dying-they-are-evolving-into-spaces-for-innovation-44820.

91. Christopher Horn, "The Green Factory: Urban forestry is ready and able to help tackle unemployment in urban communities across the U.S.," *American Forests*, Summer 2018, https://www.americanforests.org/magazine/article/the-green-factory/.

92. Ellen MacArthur Foundation, "A New Textiles Economy: Redesigning Fashion's Future," November 28, 2017, https://www.ellenmacarthurfoundation.org/publications/a-new-textiles-economy-redesigning-fashions-future.

93. U.S. Environmental Protection Agency, "Advancing Sustainable Materials Management: 2015 Fact Sheet," July 2018, https://www.epa.gov/sites/production/files/2018-07/documents/2015_smm_msw_factsheet_07242018_fnl_508_002.pdf.

94. Tellus Institute with Sound Resource Management, "More Jobs, Less Pollution: Growing the Recycling Economy in the U.S.," accessed January 8, 2019, https://www.nrdc.org/sites/default/files/glo_11111401a.pdf.

95. James Goldstein and Christi Electris, "More Jobs, Less Pollution: Growing the Recycling Economy in the U.S.," Tellus, accessed January 8, 2019, https://www.tellus.org/tellus/publication/more-jobs-less-pollution-growing-the-recycling-economy-in-the-u-s.

96. CalRecycle, "Residential Waste Stream by Material Type," data retrieved from https://www2.calrecycle.ca.gov/WasteCharacterization/ResidentialStreams?lg=207&cy=1.

97. CalRecycle, "Business Group Waste Stream by Material Type," data retrieved from https://www2.calrecycle.ca.gov/WasteCharacterization/MaterialTypeStreams?lg=207&cy=1.

98. Kickstarter, "Paper Water Bottle: We Will Help Save Our Planet," accessed January 14, 2019, https://www.kickstarter.com/projects/paperwaterbottle/paper-water-bottle-we-will-help-save-our-planet.

99. Toad and Co., "The Shipping Revolution Begins Now," accessed January 14, 2019, https://www.toadandco.com/reduce-reuse-recycle.

100. L. Lebreton, et al., "Evidence that the Great Pacific Garbage Parch is Rapidly Accumulating Plastic," *Scientific Reports*, March 2018, https://www.nature.com/articles/s41598-018-22939-w.

101.Rachel Fritts, "In the Fishing Industry, Gear Recycling is Finally Catching On," *Ensia*, April 3, 2017, https://ensia.com/features/fishing-gear-recycling/.

102. ReFed, "Roadmap to Reduce U.S. Food Waste 20%," accessed January 8, 2019, https://www.refed.com/downloads/ReFED_Report_2016.pdf.

103. Andy Fisher, "The American hunger-industrial complex: Do big businesses have a vested interest in food banks?" *Lacuna Magazine*, November 14, 2017, https://lacuna.org.uk/food-and-health/do-businesses-corporations-have-a-vested-interest-in-american-food-banks/.

104. Dana Frasz, email communication with the author, June 5, 2018.

105. ReFed, "Roadmap to Reduce U.S. Food Waste 20%," accessed January 9, 2019, https://www.refed.com/download.

106. ReFed, "Food Waste Is a Solvable Problem," accessed January 8, 2019, https://www.refed.com/solutions?sort=jobs-created.

107. Megan Langner, email to the author, August 13, 2018.

108. Kristine Wong, "The Business of Farming Against the Odds," Civil Eats, November 28, 2016, https://civileats.com/2016/11/28/kitchen-table-advisors/.

109. Robin Wall Kimmerer, *Braiding Sweetgrass: Indigenous Wisdom, Scientific Knowledge and the Teachings of Plants* (Minneapolis: Milkweed Editions, 2013), 124.

110. Gretchen Daily, Nature's Services: *Societal Dependence on Natural Ecosystems,* (Washington, D.C.: Island Press, 1997), 3-4.

111. Marin Carbon Project, "Science," accessed January 8, 2019, https://www.marincarbonproject.org/marin-carbon-project-science.

112. Ibid.

113. Susan S. Lang, "'Slow, Insidious' soil erosion threatens human health and welfare as well as the environment, Cornell study asserts," *Cornell Chronicle*, March 20, 2006, http://news.cornell.edu/stories/2006/03/slow-insidious-soil-erosion-threatens-human-health-and-welfare.

114. Enrique Gili, "Carbon Farming: California Focus on Soil to Meet Climate, Water Goals," *News Deeply,* July 31, 2017, https://www.newsdeeply.com/water/articles/2017/07/31/carbon-farming-california-focus-on-soil-to-meet-climate-water-goals.

115. Insurance Information Institute, "Facts + Statistics: Wildfires," accessed January 8, 2019, https://www.iii.org/fact-statistic/facts-statistics-wildfires.

116. American Forests, "Fire Funding Fix," accessed January 8, 2019, https://www.americanforests.org/magazine/article/action-center-fire-funding-fix/.

117. Umair Irfan, "California's wildfires are hardly "natural"—humans made them worse at every step," *Vox*, updated November 19, 2018, https://www.vox.com/2018/8/7/17661096/california-wildfires-2018-camp-woolsey-climate-change.

118. USDA Forest Service and CAL FIRE, "Record 129 Million Dead Trees in California," December 12, 2017, https://www.fs.usda.gov/Internet/FSE_DOCUMENTS/fseprd566303.pdf.

119. Umair Irfan, "California's wildfires are hardly "natural"—humans made them worse at every step," *Vox*, updated November 19, 2018, https://www.vox.com/2018/8/7/17661096/california-wildfires-2018-camp-woolsey-climate-change.

120. U.S. Forest Service, "What We Believe: Mission," accessed January 14, 2019, https://www.fs.fed.us/about-agency/what-we-believe.

121. Hannah Ettema, "What are the differences between national parks and national forests?—The National Forest System," National Forest Foundation, March 12, 2013, https://www.nationalforests.org/blog/what-are-the-differences-between-national-parks-and-national-forests.

122. United States Department of Agriculture, "The Rising Cost of Fire Operations: Effects on the Forest Service's NonFire Work," August 4, 2015, https://www.fs.fed.us/sites/default/files/2015-Fire-Budget-Report.pdf.

123. U.S. Department of Agriculture, "Secretary Perdue Applauds Fire Funding Fix in Omnibus," March 23, 2018, https://www.usda.gov/media/press-releases/2018/03/23/secretary-perdue-applauds-fire-funding-fix-omnibus.

124. Matt Enzenhouser, phone conversation with the author, February 3, 2018.

125. Camille Stevens-Rumann, phone conversation with the author, January 16, 2018.

126. Mike Albrecht, phone conversation with the author, March 1, 2018.

127. Tim Feran, "Ghost River Furniture turns bug-damaged wood into unique furnishings," *The Columbus Dispatch*, Updated October 27, 2017, http://www.dispatch.com/news/20171026/ghost-river-furniture-turns-bug-damaged-wood-into-unique-furnishings.

128. The Beck Group, "California Assessment of Wood Business Innovation Opportunities and Markets (Phase II Report: Feasibility Assessment of Potential Business Opportunities)," The National Forest Foundation, December 2015, https://www.nationalforests.org/assets/pdfs/Phase-II-Report-MASTER-1-4-16.pdf.

129. Venla Hammila, LinkedIn message to the author, March 29, 2018.

130. J. Steven Butler, "BP Macondo Well Incident: U.S. Gulf of Mexico Pollution Containment and Remediation Efforts," SlideShare, accessed January 8, 2019, https://www.slideshare.net/tontpong/httpenergyclaimsnetassetsmacondowellpcrpdf.

131. Jessica Hartogs, "Three years after BP oil spill, active clean-up ends in three states," *CBS News*, updated June 10, 2013, https://www.cbsnews.com/news/three-years-after-bp-oil-spill-active-clean-up-ends-in-three-states/.

132. Suzanne Goldenberg, "Has BP really cleaned up the Gulf oil spill?" *The Guardian*, April 13, 2011, https://www.theguardian.com/environment/2011/apr/13/deepwater-horizon-gulf-mexico-oil-spill.

133. Ron Bousso, "BP Deepwater Horizon Costs Balloon to $65 Billion," *Reuters*, January 16, 2018, https://www.reuters.com/article/us-bp-deepwaterhorizon-idUSKBN1F50NL.

134. Chesapeake Bay Foundation, "Chesapeake Wildlife," accessed January 14, 2019, http://www.cbf.org/about-the-bay/more-than-just-the-bay/chesapeake-wildlife/.

135. Rob Schnabel, phone conversation with the author, May 11, 2018.

136. "Nitrogen & Phosphorus," Chesapeake Bay Foundation, accessed January 14, 2019, http://www.cbf.org/issues/agriculture/nitrogen-phosphorus.html.

137. Whirlwind Steel, "Controlling Erosion and Runoff on Construction Sites," November 30, 2015, https://www.whirlwindsteel.com/blog/bid/407593/controlling-erosion-and-runoff-on-construction-sites.

138. Jeremy Cox, "Chesapeake blue crab rebound a 'success story',"*Delmarva Now*, May 18, 2016, https://www.delmarvanow.com/story/news/local/2016/05/18/chesapeake-blue-crab-rebound-success-story/84257578/.

139. Saves the Bay, "Greening the Bay," accessed January 14, 2019, https://savesfbay.org/wp-content/uploads/2018/10/Save-The-Bay_Greening-The-Bay.pdf.

140. Baylands Ecosystem Habitat Goals Project, "Science Update 2016," accessed January 9, 2019, https://baylandsgoals.org/science-update-2016/.

141. Restore America's Estuaries, "Jobs & Dollars: Big Returns from Coastal Habitat Restoration," accessed January 14, 2019, https://estuaries.org/jobs-a-dollars-big-returns-from-coastal-habitat-restoration.

142. David Lewis, phone conversation with the author, January 31, 2018.

143. World Wildlife Fund, "Living Planet Report 2018: Aiming Higher," accessed February 7, 2019, https://s3.amazonaws.com/wwfassets/downloads/lpr2018_summary_report_spreads.pdf.

144. Damian Carrington, "Humanity has wiped out 60% of animal populations since 1970, report finds," *The Guardian*, October 29, 2018, https://www.theguardian.com/environment/2018/oct/30/humanity-wiped-out-animals-since-1970-major-report-finds.

145. Sibley Guides, "Causes of Bird Mortality," accessed January 14, 2019, http://www.sibleyguides.com/conservation/causes-of-bird-mortality/.

146. Conor Gearin, "How to Stop a Bird-Murdering Cat," *The Atlantic*, December 9, 2015, https://www.theatlantic.com/science/archive/2015/12accessories-for-your-murderous-pet/419601/.

147. United States Department of Agriculture Animal and Plant Health Inspection Service, "Program Data Report G - 2017: Animals Dispersed / Killed or Euthanized / Removed or Destroyed / Freed," accessed January 14, 2019, https://www.aphis.usda.gov/wildlife_damage/pdr/PDR-G_Report.php?fy=2017&fld=&fld_val=.

148. Marianna Grossman, email exchange with the author, January 2019.

149. Jennifer Errick, "9 Wildlife Success Stories," National Parks Conservation Association, November 2, 2015, https://www.npca.org/articles/880-9-wildlife-success-stories.

150. Map of Life, "Richness and rarity," accessed January 14, 2019, https://mol.org/patterns/richnessrarity?taxa=birds&indicator=sr.

151. United Nations Environment Programme, "Protected Planet Report 2016," accessed January 14, 2019, https://wdpa.s3.amazonaws.com/Protected_Planet_Reports/2445%20Global%20Protected%20Planet%202016_WEB.pdf.

152. Will Sullivan, "Protecting A Little More Land Could Save A Lot More Biodiversity," PBS, July 10, 2017, https://www.pbs.org/wgbh/nova/articleprotecting-a-little-more-land-could-save-a-lot-more-biodiversity/.

153. Kendall R. Jones, et al., "The Location and Protection Status of Earth's Diminishing Marine Wilderness," *Current Biology*, date, https://www.cell.com/current-biology/fulltext/S0960-9822(18)30772-3.

154. Dave Merrill and Lauren Leatherby, "Here's How America Uses Its Land," *Bloomberg*, July 31, 2018, https://www.bloomberg.com/graphics/2018-US-land-use/.

155. National Park Service, "The Wilderness Act," accessed on February 7, 2019, https://www.nps.gov/pore/learn/management/wildernessact.htm.

156. Dave Mizajewski, "Monarch Butterfly 2018 Population Down by 14.8 Percent," National Wildlife Federation, March 7, 2018, https://blog.nwf.org/2018/03/monarch-butterfly-2018-population-down-by-14-8-percent/.

157. Peter Fimrite, "California's most famous butterfly nearing death spiral," *San Francisco Chronicle*, January 16, 2019, https://www.sfchronicle.com/news/article/California-s-most-famous-butterfly-nearing-13539657.php.

158. Kim Stanley Robinson, "Empty half the Earth of its humans. It's the only way to save the planet," *The Guardian*, March 20, 2018, https://www.theguardian.com/cities/2018/mar/20/save-the-planet-half-earth-kim-stanley-robinson.

159. Wendy Johnson, "Honorable Harvest," *Tricycle Magazine*, winter 2014, https://tricycle.org/magazine/honorable-harvest/.

160. Noel Kirkpatrick, "In 1972, a computer model predicted the end of the world—and we're on track," *Mother News Network*, September 5, 2018, https://www.mnn.com/green-tech/computers/stories/MIT-computer-model-predicted-end-world-limits-of-growth.

161. Kate Raworth, *Doughnut Economics* (White River Junction, VT: Chelsea Green Publishing, 2017), 32.

162. Ibid.

163. John F. Helliwell, Richard Layard, and Jeffrey D. Sachs, "World Happiness Report 2018," accessed January 14, 2019, https://s3.amazonaws.com/happiness-report/2018/WHR_web.pdf.

164. Ben Schiller, "How Much Money Do You Need To Be Happy? More Than Most People Are Making," *Fast Company*, February 23, 2018, retrieved from https://www.fastcompany.com/40534358/how-much-money-do-you-need-to-be-happy-less-than-most-people-are-making.

165. California Air Resources Board, "Assembly Bill 32 Overview," August 5, 2014, https://www.arb.ca.gov/cc/ab32/ab32.htm.

166. California Air Resources Board, "AB 32 Scoping Plan," accessed January 8, 2018, https://www.arb.ca.gov/cc/scopingplan/scopingplan.htm.

167. Sid Voorakkara, phone conversation with the author, July 6, 2018.

168. Intergovernmental Panel on Climate Change, "Summary for Policymakers of IPCC Special Report on Global Warming of 1.5 Degrees Celsius approved by governments," October 8, 2018, http://www.ipcc.ch/pdf/session48/pr_181008_P48_spm_en.pdf.

169. David Aldana Cohen, "Apocalyptic Climate Reporting Completely Misses the Point," *The Nation*, November 2, 2018, https://www.thenation.com/article/mainstream-media-un-climate-report-analysis/.

170. Simon Jessop, "Investors managing $32 trillion in assets call for action on climate change," *Reuters*, December 9, 2018, https://www.reuters.com/article/us-climatechange-investors/investors-managing-32-trillion-in-assets-call-for-action-on-climate-change-idUSKBN1O80TR.

171. Lindsay Meiman, "New analysis: Comptroller DiNapoli cost New Yorkers over $22 billion for refusing fossil fuel divestment," 350.org, October 4, 2018, https://350.org/press-release/dinapoli-cost-nys-22-billion-divestment/.

172. Green America, "Green America Community Investing Guide 2011," accessed January 14, 2019, https://community-wealth.org/sites/clone.community-wealth.org/files/downloads/tool-greenamerica-cmty-invest.pdf.

173. Kaiser Permanente, "Excerpts from Greening Health Care: How Hospitals Can Heal the Planet," August 12, 2014, https://share.kaiserpermanente.org/article/excerpts-from-greening-health-care-how-hospitals-can-heal-the-planet/.

174. California Energy Commission, "Frequently Asked Questions about the Electric Program Investment Charge (EPIC)," accessed February 7, 2019, https://www.energy.ca.gov/research/epic/faq.html.

175. Cutting Edge Capital, "$340,000 Raised by Composting and Recycling Worker Cooperative in Massachusetts," July 8, 2015, https://www.cuttingedgecapital.com/cerodpo/.

176. Steve Rubenstein, "That $12 parcel tax voters approved two years ago is about to revive SF Bay shorelines," April 12, 2018, *San Francisco Chronicle*, https://

www.sfchronicle.com/bayarea/article/That-12-parcel-tax-voters-approved-two-years-ago-12830206.php.

177. Devahsree Saha and Skye d'Almedia, "Finance for City Leaders, United Nations Human Settlements Programme 2017, accessed January 14, 2019," http://financeforcityleaders.unhabitat.org/handbook/part-2-designing-financial-products/chapter-7-green-municipal-bonds.

178. Milkin Institute Financial Innovations Lab, "Growing the U.S. Green Bond Market, Volume 2: Actionable Strategies and Solutions," 2018, https://www.treasurer.ca.gov/growing-the-u.s.-green-bond-mkt-vol2-final.pdf.

179. Elizabeth Daigneau, "Massachusetts Uses Popularity of Environmental Stewardship to Pad Its Bottom Line," *Governing*, July 2013, http://www.governing.com/topics/transportation-infrastructure/gov-massachusetts-green-bonds-a-first.html.

180. Melissa Malkin-Weber, "Self-Help Credit Union's Green Impact Capital," *GreenMoney Journal*, accessed February 7, 2019, https://greenmoneyjournal.com/shcu/.

181. Bennett Cohen and Cory Lowe, "Feebates: A Key to Breaking U.S. Oil Addiction," Rocky Mountain Institute, August 20, 2010, https://www.rmi.org/feebates-key-breaking-u-s-oil-addiction/.

182. Marin Carbon Project, "FAQ," accessed January 8, 2019, https://www.marin carbonproject.org/faq.

183. Jeffrey Dastin, "Helicopters pluck 42 people, 5 dogs and cat from brink of California wildfire," *Reuters*, October 11, 2017, https://www.reuters.com/article/US-california-fire-helicopters/helicopters-pluck-42-people-5-dogs-and-cat-from-brink-of-california-wildfire-idUSKBN1CG319.

184. Legislative Analyst's Office, "The 2017-18 Budget: Cap-and-Trade," February 2017, https://lao.ca.gov/reports/2017/3553/cap-and-trade-021317.pdf.

185. Mark Mykleby, Patrick Doherty, and Joel Makower, *The New Grand Strategy: Restoring America's Prosperity, Security, and Sustainability in the 21st Century*, (New York: St. Martin's Press, 2016), 149.

186. "Skoll Foundation's Definition of Social Entrepreneurship," Skoll Foundation, accessed January 14, 2019, http://skoll.org/wp-content/uploads/2017/02/Short-Version_SASE-Criteria_04082016-3.pdf.

187. Frederick Daso, "Bill Gates And Elon Musk Are Worried For Automation—But This Robotics Company Founder Embraces It," *Forbes*, December 18, 2017, https://www.forbes.com/sites/frederickdaso/2017/12/18/bill-gates-elon-musk-are-worried-about-automation-but-this-robotics-company-founder-embraces-it/#6e6488bb40f8.

188. U.S. Department of Labor, Mine Safety and Health Administration, "Coal Fatalities for 1900 Through 2018," accessed February 13, 2019, https://arlweb.msha.gov/stats/centurystats/coalstats.asp.

189. Bureau of Labor Statistics, "Databases, Tables & Calculators by Subject," February 14, 2019, https://data.bls.gov/timeseries/CES1021210001.

190. Valerie Volcovici, "Awaiting Trump's coal comeback, miners reject retraining," *Reuters*, November 1, 2017, https://www.reuters.com/article/us-trump-effect-coal-retraining-insight/awaiting-trumps-coal-comeback-miners-reject-retraining-idUSKBN1D14G0.

191. Abel Gustafson, Seth Rosenthal, Anthony Leiserowitz, Edward Maibach, John Kotcher, Matthew Ballew, and Matthew Goldberg, "The Green New Deal has Strong Bipartisan Support," Yale Program on Climate Change Communication, December 14, 2018, http://climatecommunication.yale.edu/publications/the-green-new-deal-has-strong-bipartisan-support/.

192. Thomas L. Friedman, "The Green New Deal Rises Again," *New York Times*, January 8, 2019, https://www.nytimes.com/2019/01/08/opinion/green-new-deal.html.

GRATITUDE

Grateful appreciation go out to Chris and Matthew, the dear loves of my life, for their support during this project and always. They inspire my zeal to make the world a better place. Thank you to my relatives Roger Burt, Mala Burt, Con Amore Burt, Mazie Burt, Mildred Schipper, Lee Sonne, Ginny Sonne, Margaret Sonne, and Julius Sonne who instilled in me a profound appreciation for nature.

I owe a debt of gratitude to my brilliant editor Christopher Ryan with Haiku Editorial. His questions were vital in clarifying my thinking. This book would not have been possible without his incisive edits, gentle guidance, and fierce advocacy.

Tawny Vaughan, Matthew Morse, and Kira Burt's graphic design work helped take this book to the next level. I appreciate their talent and support.

Thank you to my careful reviewers: Brian Beckon, Roger Burt, Malini Kannan, Clayton Nall, Steve Raney, and Vanessa Warheit.

Kati Kallins and Kruti Ladani demonstrated unflagging research skills that impressed me time and again as they uncovered invaluable nuggets of information.

My dear friends Diane Bailey, Mike Forster, John Hogan, Kristin Kuntz-Duriseti, Diane Meier, Leann Pereira, Micah Perlin, and Vanessa Warheit provided indefatigable support.

I would be remiss if I did not mention how much I appreciate Dan Smolen's early enthusiasm and encouragement for this project.

Thank you to the following individuals who provided ideas and guidance along the way:

Dave Barmon, John Bourgeois, Kate Vershov Downing, Nicholas Eberstadt, Ann Edminster, Jen Fedrizzi, Dana Frasz, Travis Gnehm, Marianna Grossman, Peter Gruenwoldt, Jeff Harding, Chris Heltne, R. Paul Herman, Dave Hunzicker, David Jaber, Camille Johnson, Clark Kellogg, Jeremy Kelly, Robert Kuttner, David Lewis, Mike Mielke, Neal Morris, Jeneen Nammar, Susan Palmer, Andy Perkins, Ken Pianin, Tim Reader, Judith Redmond, Matt Renner, Heidi Sanborn, Bob Scher, Antonio Scherzer, Dr. Arthur Schipper, Jigar Shah, Joe Sherlock, Jerry James Stone, Cathrine Steenstrup, Toni Symonds, Nick Szegda, Elaine Uang, Ben Warheit, Rebecca Williams, Andrew Yarrow, and Blair Zimmerman.

INDEX

CPSIA information can be obtained
at www.ICGtesting.com
Printed in the USA
LVHW051510031119
636188LV00008B/122/P

9 781935 994343